T0297863

This book presents the fundamentals of chaos theory in conservative systems, providing a systematic study of the theory of transitional states of physical systems which lie between deterministic and chaotic behaviour. The authors' treatment of transitions to chaos, the theory of stochastic layers and webs, and the numerous applications of this theory, particularly to pattern symmetry, will make the book of importance to scientists from many disciplines.

The authors begin with an introductory section covering Hamiltonian dynamics and the theory of chaos. Attention is then turned to the theory of stochastic layers and webs and to the applications of the theory. The connection between the various structural properties of the webs and the symmetry properties of patterns is investigated, including discussion of dynamic generators of patterns, hydrodynamic patterns and fluid webs. The final section of the book contains a fascinating collection of patterns in art and living nature. The authors have been meticulous in providing a detailed presentation of the material, enabling the reader to learn the necessary computational methods and to apply them in other problems. The inclusion of a significant amount of computer graphics will also be an important aid to understanding.

The book will be of importantce to graduate students and researchers in physics and mathematics who are investigating problems of chaos, irreversibility, statistical mechanics and theories of spatial patterns and symmetries. The perhaps unconventional links between chaos theory and other topics will add to the book's interest.

*Weak chaos and
quasi-regular patterns*

Cambridge Nonlinear Science Series 1

Series editors
Professor Boris Chirikov, *Budker Institute of Nuclear Physics,
Novosibirsk, Russia*
Professor Predrag Cvitanović, *Neils Bohr Institute, Copenhagen*
Professor Frank Moss, *University of Missouri-St Louis*
Professor Harry Swinney, *Center for Nonlinear Dynamics,
The University of Texas at Austin*

Titles in this series

Weak chaos and quasi-regular patterns

G. M. Zaslavsky
New York University

R. Z. Sagdeev, D. A. Usikov
University of Maryland

A. A. Chernikov
Space Research Institute, Russian Academy of Sciences

Translated from the Russian by A. R. Sagdeeva

Published by the Press Syndicate of the University of Cambridge
The Pitt Building, Trumpington Street, Cambridge CB2 1RP
40 West 20th Street, New York, NY 10011–4211, USA
10 Stamford Road, Oakleigh, Victoria 3166, Australia

First published 1991
First paperback edition 1992

British Library cataloguing in publication data
Weak chaos and quasi-regular patterns.
1. Physics. Chaotic behaviour
I. Zaslavsky, G. M.
530.1

Library of Congress cataloguing in publication data
Zaslavskiĭ, G. M. (Georgiĭ Moiseevich)
Weak chaos and quasi-regular patterns / G. M. Zasolavsky, R. Z. Sagdeev,
D. A. Usikov and A. A. Chernikov translated from the Russian
by A. R. Sagdeeva
p. cm. (Cambridge nonlinear science series; 1)
Includes bibliographical references and index.
ISBN 0 521 37317 4 (hbk) ISBN 0 521 43828 4 (pbk)
1. Chaotic behaviour in systems. 2. Hamiltonian systems.
I. Title. II. Series.
Q172.5.C45Z37 1991
003′.7—dc20   90-44511 CIP

ISBN 0 521 37317 4 hardback
ISBN 0 521 43828 4 paperback

Transferred to digital reprinting 2001

Contents

Preface

How does the onset of chaos in Hamiltonian systems occur? This is one of the key questions in the modern theory of dynamic systems. However narrowly specialist this question may seem, the answer has a bearing on almost every branch of physics, including the quantum theory.

Chaos emerges as a result of specific local instability with respect to arbitrarily small perturbations of the system's orbits. It manifests itself in certain regions of phase space and within a certain range of the system's parameters. But the most remarkable feature of chaos in the fact that it is irremovable in fairly general physical situations. What is meant is the following. Under fairly typical conditions in phase space and in the space of values of parameters there always exist such regions in which the dynamics of the system is stochastic. These regions may be arbitrarily small, nevertheless, for a certain structure of the dynamic system given by its Hamiltonian, they are irremovable at any finite values of parameters. An illuminating example of this situation is Arnold's diffusion – a universal, unlimited transport of particles along the channels of a stochastic web in systems with the number of degrees of freedom exceeding two.

As we transfer from systems totally free of stochastic dynamics to systems with chaos, we encounter small regions which are seeds of chaos. In Hamiltonian systems these are stochastic layers and stochastic webs which, being the manifestation of weak chaos in these systems, at the same time perform a certain partitioning of phase space. Thus, as we see, topological properties of phase space turn out to be closely tied to the features and topology of certain regions – the seeds of chaos.

Two of the authors began studying stochastic layers in 1966 in connection with the problem of destruction of magnetic surfaces in toroidal magnetic traps. It was very soon discovered that the result concerning the existence of irremovable stochastic layers is universal. Later, it became clear that a set of stochastic layers can merge in a connected network, forming a stochastic web. This stochastic web permeating the entire phase space plays the leading role in the problem of global stability. No wonder that these problems have found numerous applications. Among the most unexpected applications are problems connected with the existence of liquid (hydrodynamic) webs and problems connected with the symmetry of regular and almost regular patterns in condensed matter.

Now it is clear that all these questions are of much interest to a broad audience of scientists. The authors have tried here to present the theory of formation of stochastic layers and webs (the theory of weak chaos), as well as its numerous applications. We have been meticulous in presenting all technical details (which are, in fact, rather simple). This will enable the reader to learn the necessary computational methods and apply them in other problems. The inclusion of a large amount of computer graphics will help to provide a better understanding of the subject.

The book has the following structure. It is divided into four parts. Part I contains necessary information concerning Hamiltonian dynamics and the theory of chaos (for more information, refer to the book by R. Z. Sagdeev, D. A. Usikov and G. M. Zaslavsky: *Nonlinear Physics* (Harwood Academic Publishers, New York, 1988)).

Part II is called 'Dynamic order and chaos'. It considers theories of stochastic layers and webs and their applications.

Part III is called 'Spatial patterns'. As it turned out, various structural properties of the webs are connected with patterns in condensed matter and, among other things, with their symmetry properties. In this part we discuss dynamic generators of patterns, hydrodynamic patterns and liquid webs.

Finally, Part IV, a kind of miscellany, is dedicated to patterns in art and living nature (phyllotaxis).

Generally speaking, the book does not require any technical knowledge and is meant also for students (especially postgraduates) of physics, mathematics and engineering. The authors also hope that those who take a special interest in this field of knowledge and those who have made chaos their speciality will find in this book much interesting and rather unconventional information concerning the theory of chaos: the connection of the theory of chaos to the theory of quasi-crystals, to the theory of plane tilings, to chaotic rotations of a satellite, to errors of computational schemes, etc.

The authors are deeply grateful to their colleagues M. Yu. Zakharov, A. I. Neishtadt, M. Ya. Natenson, B. A. Petrovichev, who participated in obtaining some of the results discussed in the book.

Part I · General concepts

1
Hamiltonian dynamics

The equations of motion in classical physics differ considerably depending upon the subject they describe: a particle, an electromagnetic field, or a fluid. However our natural yearning for unification in the description of different phenomena has long since led to the development of universal formalisms. Among these the Lagrangian and Hamiltonian formalisms are the most advanced. This can be explained by the nature of the phenomena discussed. The popularity of each method varied at different stages in the development of physics. Throughout the whole period of advancement of relativistically invariant theories, preference was chiefly given to Lagrangian formalism (this was most conspicuous in field theory and the theory of a continuous medium). To a large extent, it was not before the generalization of the concepts of Hamiltonian formalism and introduction of Poisson's brackets that the Hamiltonian method of analysis was able to compete with the Lagrangian one.

The formation of new ideas and possibilities triggered recently by the discovery of the phenomenon of dynamic stochasticity (or simply, chaos) have brought the methods of Hamiltonian dynamics to the fore. Liouville's theorems on the conservation of phase volume and on the integrability of systems with a complete set of integrals of motion have determined both the formulation of many problems of dynamics and the methods of their study. The Hamiltonian method turned out to be of extreme importance for the theory of stability, which was advanced in this direction by Poincaré. Numerous subsequent studies have shown that Hamiltonian systems (i.e., systems which can be described by Hamiltonian equations of motion) demonstrate fundamental physical differences from other (non-Hamiltonian) systems. This chapter provides the most necessary information on Hamiltonian systems. (Note 1.1)

1

1.1 Hamiltonian systems

The state of a Hamiltonian system can be described by N generalized momenta $p \equiv (p_1, \ldots, p_N)$ and the same number N generalized coordinates $q \equiv (q_1, \ldots, q_N)$. Here N designates the number of a system's degrees of freedom. The evolution of p and q in time is determined by the equations of motion

$$\dot{p}_i = -\frac{\partial H}{\partial q_i}; \qquad \dot{q}_i = \frac{\partial H}{\partial p_i}; \qquad (i = 1, \ldots, N), \tag{1.1.1}$$

which make sense only together with the Hamiltonian

$$H = H(p, q, t). \tag{1.1.2}$$

Here a dot over a symbol stands for the time t derivatives. The Hamiltonian function (or Hamiltonian) is given in $2N$-dimensional phase space (p, q) and may also be an explicit function of time. Pairs of variables (p_i, q_i) are called canonically conjugate pairs and the equations (1.1.1) are canonical equations.

Time t can also be added to the set of the system's coordinate variables. In order to do this, the phase space of a system should be expanded by way of introduction of one more pair of canonical variables

$$p_0 = -H; \qquad q_0 = t. \tag{1.1.3}$$

Now the Hamiltonian

$$\mathcal{H} = H(p, q, q_0) + p_0 \tag{1.1.4}$$

defines the following equations of motion

$$\dot{p}_i = -\frac{\partial \mathcal{H}}{\partial q_i}; \qquad \dot{q}_i = \frac{\partial \mathcal{H}}{\partial p_i}; \qquad (i = 0, 1, \ldots, N). \tag{1.1.5}$$

For $i = 1, \ldots, N$ the system (1.1.5) leads to the familiar equations (1.1.1). For $i = 0$ in accordance with (1.1.3) and (1.1.4) we get

$$\dot{p}_0 = -\frac{\partial H(p, q, q_0)}{\partial q_0} = -\frac{\partial H(p, q, t)}{\partial t};$$
$$\tag{1.1.6}$$
$$\dot{q}_0 = \frac{\partial \mathcal{H}}{\partial p_0} = 1.$$

The first equation in (1.1.6) signifies the well-known equality

$$\frac{dH(p, q, t)}{dt} = \frac{\partial H}{\partial t} \tag{1.1.7}$$

which can be easily verified by means of the equations of motion (1.1.1). The second equation in (1.1.6) reflects the definition of time as one of the coordinates.

Thus instead of an N-dimensional system with the Hamiltonian as a function of time we may consider an $(N+1)$-dimensional system with a time-independent Hamiltonian $\mathcal{H}$. However from the definitions (1.1.3) and (1.1.4) it becomes clear that $\mathcal{H} \equiv 0$. The new momentum $p_0 = -H$ does not bear any additional information. All the properties of a dynamic system can be described in a $(2N+1)$-dimensional phase space $(p_1, \ldots, p_N, q_1, \ldots, q_N, q_0 = t)$. It is therefore sometimes said that a system with the Hamiltonian $H(p, q, t)$ has $N + 1/2$ degrees of freedom.

Now let us introduce in a phase space (p, q) an element of phase volume

$$d\Gamma = dp_1 \cdots dp_N \, dq_1 \cdots dq_N \equiv dp \, dq$$

and a phase volume

$$\Gamma = \int_S d\Gamma$$

with some hypersurface S as a boundary. Generally speaking, phase volume of a dynamic system is a function of time. However, for Hamiltonian systems phase space is conserved

$$\Gamma_{t_1} = \Gamma_{t_2} \tag{1.1.8}$$

for arbitrary moments of time t_1 and t_2 (Liouville's theorem). This means in particular that phase fluid is incompressible. The property of phase volume conservation has some profound consequences. According to one of them, among all the conceivable trajectories there are none possessing an asymptotically stable equilibrium position (either points or sets of points attracting the trajectory). To put it otherwise, Liouville's theorem rules out the existence of attractors. It is possible to make a similar statement concerning the absence of repellers – the repellent points or sets of points.

The conservation of phase volume holds true not only for Hamiltonian systems. The progress of contemporary nonlinear analysis has led to a generalization of the concept of Hamiltonian systems (see [4]). Let us assume that z_i is a coordinate in $2N$-dimensional phase space, the variables z_i not yet separated into generalized coordinates and generalized momenta. First, the operation of generalized Poisson's brackets should be introduced. For the two arbitrary functions $A(z)$ and $B(z)$, Poisson's brackets are defined by the formula

$$[A, B] = \sum_{i,k=1}^{2N} g_{ik} \frac{\partial A}{\partial z_i} \frac{\partial B}{\partial z_k} \tag{1.1.9}$$

where the tensor $g_{ik} = g_{ik}(z)$ in general depends on the variables. It is also assumed that the following conditions are satisfied:

   (a) the bilinearity condition

$$[aA + bB, C] = a[A, C] + b[B, C] \qquad (1.1.10)$$

where a and b are constants;

   (b) the skew-symmetry condition

$$[A, B] = -[B, A]. \qquad (1.1.11)$$

This condition, for example, cannot be satisfied if the phase space dimensionality is uneven;

   (c) Leibniz equality

$$[AB, C] = B[A, C] + A[B, C]; \qquad (1.1.12)$$

   (d) Jacoby's equality

$$[A, [B, C]] + [C, [A, B]] + [B, [C, A]] = 0. \qquad (1.1.13)$$

With the help of Poisson's brackets (1.1.9) we are able to define the tensor g_{ik} as follows

$$g_{ik} = [z_i, z_k], \qquad (1.1.14)$$

while the brackets themselves can be expressed in the following way

$$[A, B] = \sum_{i,k=1}^{2N} \frac{\partial A}{\partial z_i} \frac{\partial B}{\partial z_k} [z_i, z_k]. \qquad (1.1.15)$$

Now let us consider the arbitrary function $H = H(z)$, which is referred to as the Hamiltonian. A system is called a generalized Hamiltonian system if it can be described by the following equations of motion

$$\dot{z}_i = [z_i, H], \qquad (i = 1, \ldots, 2N). \qquad (1.1.16)$$

The variations of any system $A(z)$ with time can be defined with the help of the equations (1.1.14)-(1.1.16):

$$\dot{A} = \sum_{i=1}^{2N} \frac{\partial A}{\partial z_i} \dot{z}_i = \sum_{i=1}^{2N} \frac{\partial A}{\partial z_i} [z_i, H] = [A, H]. \qquad (1.1.17)$$

Specifically, if g_{ik} is an identity skew-symmetry matrix,

$$g_{ik} = \begin{pmatrix} 0 & 1 \\ -1 & 0 \end{pmatrix},$$

equations (1.1.16) are equivalent to equations (1.1.1) for $N = 1$ and $z_1 = q$, $z_2 = p$. For an arbitrary N, the equations (1.1.1) should follow from (1.1.16), if we assume that

$$g_{ik} = \begin{pmatrix} 0 & \hat{1} \\ -\hat{1} & 0 \end{pmatrix}, \qquad (1.1.18)$$

where $\hat{1}$ is the identity matrix of the order of N and

$$(z_1, \ldots, z_n) = (q_1, \ldots, q_N);$$

$$(z_{N+1}, \ldots, z_{2N}) = (p_1, \ldots, p_N).$$

The above generalized form of Hamiltonian dynamics will be applied later during an analysis of equations of motion of vector fields.

In the special case of (1.1.18), where the Hamiltonian equations (1.1.1) are true, Poisson's brackets may be presented in the following way

$$[A, B] = \sum_{i=1}^{N} \left(\frac{\partial A}{\partial q_i} \frac{\partial B}{\partial p_i} - \frac{\partial B}{\partial q_i} \frac{\partial A}{\partial p_i} \right) \qquad (1.1.19)$$

and time derivatives (1.1.6) can be presented as

$$\dot{A} = \sum_{i=1}^{N} \left(\frac{\partial A}{\partial q_i} \frac{\partial H}{\partial p_i} - \frac{\partial H}{\partial q_i} \frac{\partial A}{\partial p_i} \right). \qquad (1.1.20)$$

The equations of motion written in the form (1.1.1) or (1.1.16), or equivalent definitions of Poisson's brackets, are fundamental to canonical or Hamiltonian formalism.

1.2 The phase portrait

The system's family of trajectories in phase space comprises its phase portrait. The simplest form of the phase portrait is obtained for $N = 1$. Phase space then is a plane (p, q). One-dimensional motion, for example, may be defined by the Hamiltonian

$$H = \frac{1}{2m} p^2 + V(q), \qquad (1.2.1)$$

where $V(q)$ is the potential energy (the potential) of a particle. According to (1.1.7), if the Hamiltonian is not an explicit function of time, then $\dot{H} = 0$ and H is an integral of motion (invariant):

$$H = \text{const} = E, \qquad (1.2.2)$$

where E is some fixed value of the invariant. In the case of (1.2.1) the quantity E is the system's total energy. In the example (1.2.1), motion occurs on the surface of constant energy and

$$p = \pm\{2m[E - V(q)]\}^{1/2} = m\dot{q}. \tag{1.2.3}$$

Equation (1.2.3) defines a parametric family of trajectories comprising the phase portrait of the system (Fig. 1.2.1).

If a system's trajectory is localized in a finite region of phase space, the corresponding motion is said to be finite. Otherwise, it is infinite.

Singularities in phase space are defined as the fixed points of the equations of motion, so that they may be derived from the following equations

$$\frac{\partial H}{\partial q_i} = 0; \qquad \frac{\partial H}{\partial p_i} = 0. \tag{1.2.4}$$

Fig. 1.2.1 Potential $V(q)$ (a) and the corresponding phase portrait (b) of one-dimensional motion. The trajectories C_1 and C_2 are separatrices.

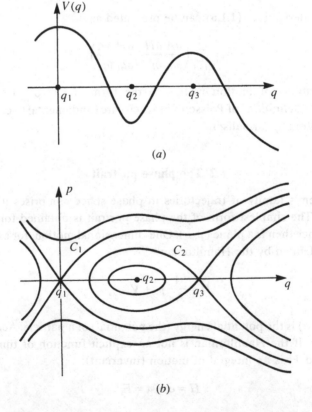

(a)

(b)

In case (1.2.1), these are the points where $p = 0$ and the potential $V(q)$ has an extremum: $V'(q) = 0$.

Figure 1.2.1 shows that these points can be either of the elliptic type (q_2), or of the hyperbolic type (q_1), or saddles. In the neighbourhood of an elliptic point the motion is stable, and the trajectories have the shape of ellipses (Fig. 1.2.2a). In the neighbourhood of a saddle the motion is unstable, and the trajectories have the shape of hyperbolas (Fig. 1.2.2b). The trajectory passing through a saddle is called a separatrix (trajectories C_1 and C_2 on Fig. 1.2.1). A saddle always has entering and outgoing whiskers of separatrices. (1.2.2b).

The absence of limit points and limit quantities such as a limit cycle makes the phase portrait of Hamiltonian systems less diverse. Nevertheless, it does not lessen their complexity. Introduction of dissipative factors and the asymptotically limited trajectories associated with them, makes the system in a way less sensitive to various small perturbations (Note 1.2). This book is almost exclusively dedicated to the properties of the Hamiltonian systems, which are absent in the dissipative case (Note 1.3). Therefore, having made this remark, we do not go into any further details here.

1.3 'Action-angle' variables

For the sake of convenience, we shall first introduce 'action-angle' variables for the case of motion with one degree of freedom (1.2.1). In the case of finite and therefore periodic motion, action I is defined by the expression:

$$I = \frac{1}{2\pi} \oint p \, dq, \qquad (1.3.1)$$

where integration is performed over a closed orbit of the system. In the

Fig. 1.2.2 The trajectories in the vicinity of an elliptic (a) and a hyperbolic point (b).

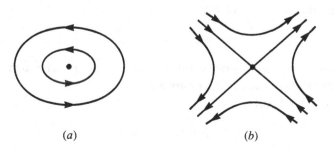

(a)   (b)

case of infinite but still periodic motion, integration in (1.3.1) should be performed over the whole period. The value

$$S(q, I) = \int^q p(q, H)\, dq \qquad (1.3.2)$$

is called truncated action. Both here and in (1.3.1) the expression for $p = p(q, H)$ is given in equation (1.2.3). We should also substitute (1.3.1) for I in (1.3.2) having replaced H by $H(I)$. In this case the value $S(q, I)$ is also a generating function. It enables us to define a new coordinate, the angle:

$$\theta = \frac{\partial S(q, I)}{\partial I}. \qquad (1.3.3)$$

The variables (I, θ) make up a canonically conjugate pair, i.e. the Hamiltonian equations of motion will be valid for them:

$$\dot{I} = -\frac{\partial H(I)}{\partial \theta} = 0;$$

$$\dot{\theta} = \frac{\partial H(I)}{\partial I} = \frac{dH(I)}{dI} \equiv \omega(I). \qquad (1.3.4)$$

Being expressed as a function of the integral of motion H, the action is itself an integral. This is reflected in the first equation in (1.3.4). The second equation defines the frequency of periodic motion $\omega(I)$. In general, it is the function of action and the system's energy $E = H$. The dimensionless parameter

$$\alpha = \frac{I}{\omega}\frac{d\omega}{dI} \qquad (1.3.5)$$

determines the degree of this dependence. If $\alpha \neq 0$, the oscillations are called nonlinear. Their frequency is a function of energy. Now we can integrate the equations of motion (1.3.4):

$$I = \text{const}; \qquad \theta = \omega(I)t + \theta_0. \qquad (1.3.6)$$

That is why the 'action-angle' variables are so useful (although, as we shall see later, this is not their only merit).

For a linear oscillator:

$$H = \tfrac{1}{2}p^2 + \tfrac{1}{2}\omega_0^2 q^2 \qquad (1.3.6a)$$

(unit mass is assumed). With the help of definitions (1.3.1)–(1.3.3) we can easily obtain

$$q = (2I/\omega)^{1/2} \cos \theta;$$

$$p = (2I\omega)^{1/2} \sin \theta; \qquad (1.3.7)$$

$$\theta = \omega_0 t.$$

From (1.3.6a) and (1.3.7) there follows

$$H = \omega_0 I, \qquad (1.3.8)$$

i.e., $\omega = \text{const} = \omega_0$ and $\alpha = 0$. This is exactly what determines the linearity of the oscillator (1.3.6).

Applying the definitions of variables (I, θ) we can express the old variables (p, q) in terms of the new ones

$$p = p(I, \theta); \qquad q = q(I, \theta). \qquad (1.3.9)$$

Due to the cyclicity of the variable θ, i.e. phase shift in θ by $2\pi n$, for an integer n, does not change the expression (1.3.9), q and p can be expanded in a Fourier series:

$$q = q(I, \theta) = \sum_{n=-\infty}^{\infty} a_n(I) \, e^{in\theta};$$

$$(1.3.10)$$

$$p = p(I, \theta) = \sum_{n=-\infty}^{\infty} b_n(I) \, e^{in\theta},$$

where the coefficients a_n and b_n are equal to:

$$a_n = \frac{1}{2\pi} \int_0^{2\pi} d\theta \, e^{-in\theta} q(I, \theta);$$

$$(1.3.11)$$

$$b_n = \frac{1}{2\pi} \int_0^{2\pi} d\theta \, e^{-in\theta} p(I, \theta).$$

The Fourier harmonics (1.3.11) determine the spectral properties of the system. As the variables (p, q) are real, the following relations for the expansion coefficients can be presented:

$$a_{-n} = a_n^*; \qquad b_{-n} = b_n^*. \qquad (1.3.12)$$

Out of general considerations, the important asymptotic property of the system's spectrum can be found [6] if the system motion is periodic:

$$a_n \sim \exp(-n/N_0);$$

$$b_n \sim \exp(-n/N_0) \qquad (n \to \infty), \qquad (1.3.13)$$

where N_0 is a constant, defining an effective number of harmonics within the spectrum of the system's oscillations. For $n > N_0$ the amplitudes of the Fourier harmonics are exponentially small. Specifically, in the case of a linear oscillator, the oscillations (1.3.7) contain exactly one harmonic.

1.4 The nonlinear pendulum

The nonlinear pendulum is a common physical model. This is due to the fact that many problems concerning oscillations can be more or less easily reduced to the equations of the nonlinear pendulum. Its Hamiltonian has the form

$$H = \tfrac{1}{2}\dot{x}^2 - \omega_0^2 \cos x, \qquad (1.4.1)$$

where unit mass is assumed, i.e. $p = \dot{x}$, and ω_0 is the frequency of weak oscillations. The equation of motion for a nonlinear pendulum is as follows:

$$\ddot{x} + \omega_0^2 \sin x = 0 \qquad (1.4.2)$$

and its phase portrait is as shown in Fig. 1.4.1. The singularities are of the elliptic type ($\dot{x} = 0$, $x = 2\pi n$), and saddles ($\dot{x} = 0$, $x = 2\pi(n+1)$); $n = 0, \pm 1, \ldots$. When $H < \omega_0^2$, the trajectories correspond to the pendulum's oscillations (finite motion), in the case of $H > \omega_0^2$, to the pendulum's rotation (infinite motion). Trajectories with $H = \omega_0^2$ are separatrices. The solution on a separatrix can be obtained if we substitute

$$H = H_s = \omega_0^2 \qquad (1.4.3)$$

in the equation (1.4.1). This gives us the following equation:

$$\dot{x} = \pm 2\omega_0 \cos(x/2). \qquad (1.4.4)$$

Having supplied the initial condition: $t = 0$, $x = 0$, we get the solution in the form:

$$x = 4 \arctan \exp(\pm \omega_0 t) - \pi. \qquad (1.4.5)$$

Two whiskers of a separatrix (one entering the saddle and another leaving it) correspond to the different signs of t. With the help of (1.4.4) we can obtain from (1.4.5)

$$v = \dot{x} = \pm \frac{2\omega_0}{\cosh \omega_0 t}. \qquad (1.4.6)$$

The solution of this form is called a soliton.

Now let us take a closer look at the method of finding a solution for equation (1.4.1) (Note 1.4). Let us introduce a new parameter

$$x^2 = \frac{\omega_0^2 + H}{2\omega_0^2} = \frac{H + H_s}{2H_s} \qquad (1.4.7)$$

defining dimensionless energy with another initial point. The action can be obtained with the help of the formula (1.3.1)

$$I = I(H) = \frac{8}{\pi} \omega_0 \begin{cases} E(\pi/2; x) - (1 - x^2)F(\pi/2; x), & (x \leqslant 1) \\ x E(\pi/2; 1/x), & (x \geqslant 1), \end{cases} \qquad (1.4.8)$$

where $F(\pi/2; x)$ and $E(\pi/2; x)$ are complete elliptic integrals of the first and second kind, respectively. From the definition of the frequency

Fig. 1.4.1 The periodic potential of a pendulum (a) and its phase plane (b).

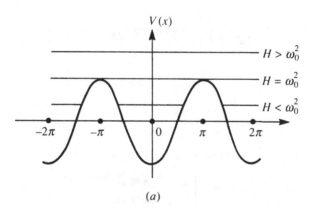

(a)

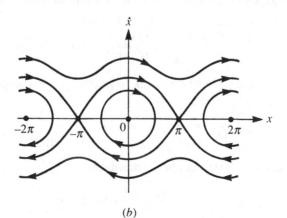

(b)

of oscillations (1.3.4) and from (1.4.8), it follows that

$$\omega(H) = \frac{dH(I)}{dI} = \left[\frac{dI(H)}{dH} \right]^{-1}$$

$$= \frac{\pi}{2} \, \omega_0 \begin{cases} 1/F(\pi/2; \varkappa), & (\varkappa \leqslant 1) \\ \varkappa/F(\pi/2; 1/\varkappa), & (\varkappa \geqslant 1). \end{cases} \tag{1.4.9}$$

The solution of the equation of motion (1.4.2) for velocity $\dot{x}$ has the form

$$\dot{x} = 2\varkappa\omega_0 \begin{cases} \mathrm{cn}(t; \varkappa), & (\varkappa \leqslant 1) \\ \mathrm{dn}(t; \varkappa), & (\varkappa \geqslant 1), \end{cases} \tag{1.4.10}$$

where cn and dn are Jacobian elliptic functions. For $\varkappa = 1$, the expressions (1.4.10) take the form of (1.4.6). Expansion of (1.4.10) into a Fourier series yields [8]:

$$\dot{x} = 8\omega \sum_{n=1}^{\infty} \frac{a^{n-1/2}}{1 + a^{2n-1}} \cos[(2n-1)\omega t], \qquad (\varkappa \leqslant 1)$$

$$\dot{x} = 8\omega \left\{ \sum_{n=1}^{\infty} \frac{a^n}{1 + a^{2n}} \cos n\omega t + \frac{1}{4} \right\}, \qquad (\varkappa \geqslant 1), \tag{1.4.11}$$

where

$$a = \exp(-\pi F'/F)$$
$$\omega = \omega(H)$$
$$F = F(\pi/2; \bar{\varkappa});$$
$$F' = F[\pi/2; (1 - \bar{\varkappa}^2)^{1/2}]$$
$$\bar{\varkappa} = \begin{cases} \varkappa, & (\varkappa \leqslant 1) \\ 1/\varkappa, & (\varkappa \geqslant 1). \end{cases} \tag{1.4.12}$$

Using the asymptotic expansions of the elliptic integrals we get

$$F(\pi/2; \varkappa) \sim \begin{cases} \pi/2, & (\varkappa \ll 1) \\ \frac{1}{2} \ln \dfrac{32 H_s}{H_s - H}, & (1 - \varkappa^2 \ll 1). \end{cases} \tag{1.4.13}$$

Using these equations (1.4.12) the asymptotics for the parameter of expansion a can be found,

$$a \sim \begin{cases} \varkappa^2/32, & (\varkappa \ll 1) \\ \exp(-\pi/N_0), & (1 - \varkappa^2 \ll 1), \end{cases} \tag{1.4.14}$$

where a new parameter has been introduced:

$$N_0 = \frac{\omega_0}{\omega(H)} = \frac{2}{\pi} F(\pi/2; \varkappa) \tag{1.4.15}$$

which determines the ratio between the frequency of the pendulum's small oscillations and its frequency at a given energy H. Its asymptotic expansions according to (1.4.13), are

$$N_0 \sim \begin{cases} 1, & (\varkappa \ll 1) \\ \dfrac{1}{\pi} \ln \dfrac{32H_s}{H_s - H}, & (1 - \varkappa^2 \ll 1). \end{cases} \qquad (1.4.16)$$

From (1.4.14) it follows that the number N_0 determines an effective number of harmonics in the Fourier spectrum. If $n > N_0$ the amplitudes decrease exponentially. Otherwise, if $n < N_0$ then all the coefficients of the Fourier transform are of the same order of magnitude.

The above gives us the full notion of the character of the pendulum's oscillations. The case of $\varkappa \ll 1$ corresponds to the small amplitudes of oscillations, while H is close to $-H_s$ (i.e., the oscillations are taking place near the bottom of the potential well (see Fig. 1.4.1)). In accordance with (1.4.14) and (1.4.16), in this case $\omega(H) \approx \omega_0$, $N_0 \sim 1$ and the amplitudes of a are small. Consequently, in equations (1.4.11) we need to retain only the first term, which corresponds to the linear oscillations

$$v = \dot{x} \approx (\delta H)^{1/2} \cos \omega_0 t,$$

where

$$\delta H = H_s + H = \omega_0^2 - |H|.$$

Let us consider the case of $\varkappa^2 \to 1$, i.e. $H \to H_s$. In the vicinity of a separatrix, the frequency $\omega(H) \to 0$ and the period of oscillation diverges logarithmically. The pendulum's velocity $\dot{x}$, as a function of time, approximates a periodic sequence of soliton-like pulses (Fig. 1.4.2). The distance between two adjacent crests in the same phase hardly differs from the period of oscillation $2\pi/\omega(H)$, and the width of each pulse is close to $2\pi/\omega_0$. The number N_0 thus also determines the ratio of distance between

Fig. 1.4.2 The dependence of a pendulum's velocity on time near a separatrix.

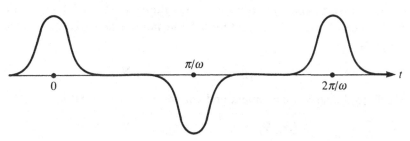

the soliton-like pulses to their width for the oscillations in the neighbour-hood of a separatrix.

The described properties of oscillations hold true not only when we are approaching the separatrix from below $(H \to \omega_0^2 - 0)$, but also for $H \to \omega_0^2 + 0$, i.e., when we are approaching it from the side of the infinite trajectories.

1.5 Multidimensional motion

Here we shall be considering multidimensional integrable systems. The integral of motion of a system

$$H = H(p, q) = H(p_1, q_1; \ldots; p_N, q_N)$$

is the quantity $\mathscr{F}(p, q)$, commuting with the Hamiltonian

$$[H, \mathscr{F}] \equiv 0,$$

where the commutator, or Poisson's brackets, operation is defined according to (1.1.19) as

$$[A, B] \equiv \sum_{i=1}^{N} \left(\frac{\partial A}{\partial p_i} \frac{\partial B}{\partial q_i} - \frac{\partial A}{\partial q_i} \frac{\partial B}{\partial p_i} \right). \tag{1.5.1}$$

Let us henceforth always assume, with the exception of especially mentioned cases, that various values of the energy E of the system correspond to the surfaces of constant values of the Hamiltonian

$$H(p, q) = \text{const} = E.$$

An integral of motion $\mathscr{F}$, which makes it possible to reduce the order of a set of equations, is called the first integral. In the general case, to integrate a set of differential equations of the order of $2N$, we need the same number of first integrals. However, in the case of the Hamiltonian equations of motion, N first integrals are sufficient. This fact is reflected by the Liouville–Arnold theorem (Note 1.5).

Let us assume that we are given a Hamiltonian system $H(p, q)$ with N degrees of freedom, executing a finite motion and having N first integrals

$$\mathscr{F}_i = \mathscr{F}_i(p, q), \qquad (i = 1, \ldots, N)$$

which are linearly independent and commute pairwise, that is

$$[\mathscr{F}_i, \mathscr{F}_j] \equiv 0, \qquad (i, j = q, \ldots, N).$$

Then:

(i) the trajectories of the system lie on the surface of an N-dimensional torus;

(ii) the motion is conditionally periodic and is characterized by N frequencies

$$\omega_i = \omega_i(\mathscr{F}_1, \ldots, \mathscr{F}_N), \qquad (i = 1, \ldots, N);$$

(iii) the angular variables θ_i, which determine the coordinates on the surface of the torus satisfy the equations

$$\dot{\theta}_i = \omega_i(\mathscr{F}_1, \ldots, \mathscr{F}_N), \qquad (i = 1, \ldots, N) \qquad (1.5.2)$$

which can be integrated at once and yield:

$$\theta_i = \omega_i t + \text{const}_i, \qquad (i = 1, \ldots, N). \qquad (1.5.3)$$

A sample trajectory is shown in Fig. 1.5.1.

In the case of two degrees of freedom, only one additional first integral, independent of H, is necessary.

During the motion, the trajectory does not leave the surface of the torus. Therefore, the torus is the system's invariant characteristic. A set of invariant tori corresponds to the set of different values of integrals $(\mathscr{F}_1, \ldots, \mathscr{F}_N)$. Their relative position in phase space is determined by its dimensionality. For $N = 2$ the tori corresponding to the different values of the integrals $(\mathscr{F}_1, \mathscr{F}_2)$ are nested in one another and do not intersect (Fig. 1.5.2). In this case it is said that the tori divide the space. For $N > 2$ the tori intersect and do not divide the space. This is easily explained in the following way.

In $2N$-dimensional phase space, the surface of constant energy $H = E$ has a dimensionality of $2N - 1$. The borders dividing it into different regions have dimensionality of $2N - 2$. If the tori divide the space then their dimensionality N must satisfy the following condition

$$N \geqslant 2N - 2.$$

Fig. 1.5.1 The trajectory on a two-dimensional torus.

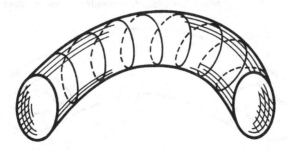

Hence $N \le 2$. Thus, the case of two degrees of freedom stands in a class by itself. Later we shall see how this topological singularity of phase space affects the dynamic stability of the system.

The motion on the surface of a two-dimensional torus is characterized by the two frequencies ω_1 and ω_2. In the general case, they are incommensurate and the trajectory densely covers the entire surface of the torus and is open. The case where the ratio between the frequencies is the rational number

$$\frac{\omega_1}{\omega_2} = \frac{m_1}{m_2} \tag{1.5.4}$$

(m_1 and m_2 are integers), is said to be one of resonance. The trajectory closes on itself after a finite number of revolutions around the torus.

In the general case, the condition of resonance has the following form:

$$\sum_{i=1}^{N} m_i \omega_i = 0. \tag{1.5.5}$$

Here m_i are integers, among which at least one must be non-zero.

Since the frequencies ω_i are functions of the integrals of motion $\mathscr{F}_i$, the conditions of resonance (1.5.5) imply a certain relation among $\mathscr{F}_i$. Therefore, resonance is always accompanied by a certain degree of degeneracy of the system. Generally speaking, the integrals $\mathscr{F}_i$ and the angles θ_i do not form canonical pairs of variables. In the case of an integrable system, canonical variables are defined in the following way. Since the motion occurs on the surface of an N-dimensional torus, it is possible to choose exactly N basis contours C_j, which can be neither contracted to a point, nor transformed into one another. The basis contours for the case of $N=2$ are presented in Fig. 1.5.3. By means of the contours C_j, we can now define N independent action variables as:

$$I_j = \frac{1}{2\pi} \oint_{C_j} \sum_{i=1}^{N} p_i \, dq_i, \quad (j=1, \ldots, N), \tag{1.5.6}$$

Fig. 1.5.2   At $N=2$, invariant tori divide the phase space.

which are the integrals of motion (the so-called first integral invariants of Poincaré [2]).

The integrals $\mathscr{F}_i$, being linearly independent, can be expressed in terms of the actions I:

$$\mathscr{F}_i = \mathscr{F}_i(I_1, \ldots, I_N), \qquad (i = 1, \ldots, N).$$

Similarly, the Hamiltonian H is also expressed in terms of actions:

$$H = H(I_1, \ldots, I_N)$$

i.e. the variables θ_i are cyclic. The canonical equations now have the following form:

$$\dot{I}_i = -\frac{\partial H}{\partial \theta_i}; \qquad \dot{\theta}_i = \frac{\partial H}{\partial I_i}, \qquad (i = 1, \ldots, N). \tag{1.5.7}$$

Hence

$$I_i = \text{const}; \qquad \theta_i = \omega_i t + \text{const}, \qquad (i = 1, \ldots, N) \tag{1.5.8}$$

where

$$\omega_i = \omega_i(I_1, \ldots, I_N) = \frac{\partial H}{\partial \omega_i}, \qquad (i = 1, \ldots, N). \tag{1.5.9}$$

We choose the basis contours C_j somewhat arbitrarily. Therefore, the definition of action (1.5.6) is not unique. However, if the system is nondegenerate, i.e.,

$$\det \left| \frac{\partial \omega_i(I)}{\partial I_j} \right| = \det \left| \frac{\partial^2 H(I)}{\partial I_i \, \partial I_j} \right| \neq 0, \tag{1.5.10}$$

Fig. 1.5.3 Basis contours C_1 and C_2 on a two-dimensional torus.

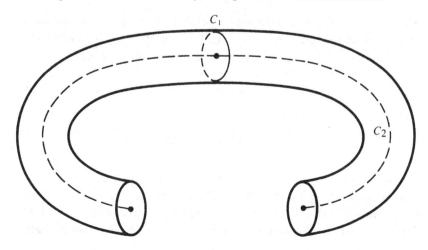

then the invariant tori

$$(I) \equiv (I_1, \ldots, I_N) = \text{const}$$

are defined unequally. According to the formulas (1.5.9), the condition (1.5.10) implies that N frequencies ω_i are independent.

The formulas (1.5.8) enable us to clarify the question of the spectral properties of the system's trajectory. To accomplish this, one only has to know the relationship between the variables (p, q) and the 'action-angle' variables (I, θ):

$$p = p(I, \theta); \qquad q = q(I, \theta).$$

Using the cyclical nature of the variable θ, we get:

$$q = \sum_{m=-\infty}^{\infty} q_m(I)\, e^{im\theta}$$

$$p = \sum_{m=-\infty}^{\infty} p_m(I)\, e^{im\theta}, \tag{1.5.11}$$

where the Fourier coefficients satisfy the condition that q and p must be real:

$$q_m^* = q_{-m}; \qquad p_m^* = p_{-m}$$

and the vectors m and θ are equal to

$$m = (m_1, \ldots, m_N)$$

$$\theta = (\theta_1, \ldots, \theta_N).$$

The numbers m_j are integers, either positive or negative. Summation in (1.5.11) is performed over all the numbers m_j. With the help of (1.5.8), the expansion (1.5.11) can be presented in the following form:

$$q = \sum_{m=-\infty}^{\infty} \tilde{q}_m(I)\, e^{im\omega t}$$

$$p = \sum_{m=-\infty}^{\infty} \tilde{p}_m(I)\, e^{im\omega t}, \tag{1.5.12}$$

where $\omega = (\omega_1, \ldots, \omega_N)$ and the definitions of amplitudes $\tilde{q}_m$ and $\tilde{p}_m$ contain constant initial phases. If there is no degeneracy, i.e., if the condition (1.5.10) is satisfied, the expansion (1.5.12) describes an N-frequency motion. This kind of motion is said to be conditionally periodic.

1.6 The Poincaré mapping

In many cases, it is convenient to define the dynamic evolution of a system, not with the help of the continuous functions $p(t)$, $q(t)$, but in a certain discrete way. Suppose that we are given a sequence of time instants $(t_0, t_1, \ldots)$. At each of these instants t_i there is determined a pair of numbers (p_i, q_i)

$$p_i = p(t_i); \qquad q_i = q(t_i)$$

lying on the system's trajectory. The following relation

$$(p_{n+1}, q_{n+1}) = \hat{T}_n(p_n, q_n) \tag{1.6.1}$$

defines the operator of displacement over a time $\hat{T}_n$. It defines the mapping of phase space onto itself and is equivalent to the equations of motion in the differential form.

Let, for example, the Hamiltonian of a system be a periodic function of time

$$H(p, q; t + T) = H(p, q; t)$$

where T is the period. Then it is sufficient to consider the solutions of the equations of motion on an interval $(t_0, t_0 + T)$ and find the form of operator $\hat{T}$ by way of matching of the solutions of two adjacent intervals

$$(t_0, t_0 + T); \qquad (t_0 + T, t_0 + 2T).$$

The finding of a solution now is reduced to a step-by-step iterative process

$$(p_n, q_n) = \hat{T}^n(p_0, q_0), \tag{1.6.2}$$

where (p_0, q_0) is the initial condition. The trajectory of a system is determined by means of (1.6.1) and (1.6.2) and is a discrete sequence of points (p_0, q_0); (p_1, q_1); $\ldots$; (p_n, q_n); $\ldots$.

In the case of $H = H(p, q, t)$, when p and q are the only variables ($N = 3/2$), the mapping (1.6.1) transforms a plane into another plane. One can apply this method also in the case of $N \geq 2$, when it is practically impossible to plot trajectories in phase space of dimensionality ≥ 4. Let us consider a plane in phase space, marking the points where the trajectory passes through this plane in a certain direction (Fig. 1.16.1). If (p_n, q_n) is a point on the plane corresponding to the nth intersection with the plane, the relation (1.6.1) defines the Poincaré mapping. The sequence of time instants marking the intersection with the plane $(t_0, t_1, \ldots)$, is not equidistant.

The Poincaré mapping acts in a space of a smaller dimensionality than that of phase space. This simplification is paid for by the effort necessary to find the form of the operator $\hat{T}_n$.

If a system has an invariant torus, its intersection with a plane, obviously, produces an invariant curve. Indeed, a trajectory everywhere densely covering the surface of a torus passes through the plane in the points everywhere densely covering the curve resulting from the intersection of the torus with the plane.

For many cases, Poincaré mapping makes the analysis of the problem less difficult. It also enables a graphic presentation of a dynamic picture in its numerical analysis. Later, these statements will be discussed in more detail.

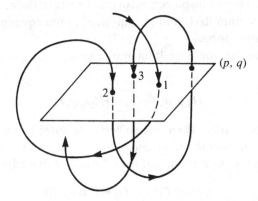

Fig. 1.6.1   Consecutive points of the Poincaré mapping.

2
Stability and chaos

Situations in which the system's integrability is immediately evident are very rare. Far more often, one has to do with a problem where a small perturbation is added to an otherwise integrable system. A large variety of methods and developed techniques exist to help analyse the effect of a perturbation on a system. Each is, in one way or another, related to the problem of a system's stability. Here one should remember that both the formulation of the stability problem and the very concept of stability are highly conditional. Any particular question concerning the system's properties determines not only the method of solution, but a specific definition of stability as well. For instance, the small perturbations may result in a weak response of the system's parameters either during a certain finite time interval or during an infinite time. In the latter case, the phase space topology of the system may either remain unchanged or may change considerably, etc.

Recently, studies on the stability of dynamic systems have been enriched by new methods and concepts. This is due to the discovery of a new phenomenon – dynamic chaos (or, simply, chaos). It turns out that while the equations may show no trace of random sources, the trajectories of a system can be the result of random time-dependent processes. The property of chaos characterizes the system as non-integrable. It is inherent only in nonlinear systems.

In this chapter we shall provide the most essential information on stability and chaos in dynamic systems. (Note 2.1)

2.1 Nonlinear resonance

Nonlinear resonance is an essential property of dynamic systems. It characterizes their response to an external perturbation and helps to understand the role played by nonlinearity.

We shall describe the unperturbed motion of the system with one degree of freedom with the Hamiltonian $H_0(I)$ and the perturbed motion with the Hamiltonian

$$H = H_0(I) + \varepsilon V(I, \theta, t) \tag{2.1.1}$$

where ε is a small dimensionless parameter ($\varepsilon \ll 1$). The perturbation V is periodic in time with the period

$$T = 2\pi/\nu.$$

Therefore, V can be expanded in a double Fourier series in θ and t:

$$V(I, \theta, t) = \tfrac{1}{2} \sum_{k,l} V_{k,l}(I)\, e^{i(k\theta - l\nu t)} + \text{c.c.}$$

$$V_{k,l}^* = V_{-k,-l}, \tag{2.1.2}$$

where c.c. designates the complex conjugate terms.

Making use of expressions (2.1.1) and (2.1.2), the canonical equations of motion can be transformed into the following:

$$\dot{I} = -\varepsilon \frac{\partial V}{\partial \theta} = -\tfrac{1}{2}\varepsilon \sum_{k,l} k V_{k,l}(I)\, e^{i(k\theta - l\nu t)} + \text{c.c.}$$

$$\dot{\theta} = \frac{\partial H}{\partial I} = \frac{dH_0}{dI} + \varepsilon \frac{\partial V(I, \theta, t)}{\partial I} \tag{2.1.3}$$

$$= \omega(I) + \tfrac{1}{2}\varepsilon \sum_{k,l} \frac{\partial V_{k,l}(I)}{\partial I}\, e^{i(k\theta - l\nu t)} + \text{c.c.}$$

Here the frequency of oscillations

$$\omega(I) = \frac{dH_0}{dI}$$

was introduced for unperturbed motion.

The resonance condition for a system implies that the following equation must be satisfied:

$$k\omega(I) - l\nu = 0. \tag{2.1.4}$$

This means that we have to specify a pair of integers (k_0, l_0), for which the corresponding I_0 will transform (2.1.4) into the following identity:

$$k_0 \omega(I_0) = l_0 \nu. \tag{2.1.5}$$

As a rule, such values $(k_0, l_0; I_0)$ can be found in abundance. This is largely due to the system's nonlinearity, i.e., to the $\omega(I)$ dependency.

The way to deal with equations (2.1.3) is to analyse certain simplified situations. Firstly, we should examine an isolated resonance (2.1.5), ignoring all other possible resonances. This actually means that in the equations of motion (2.3.1), only those terms are to be retained ($k = k_0$, $l = l_0$), which satisfy the resonance condition at $I = I_0$:

$$\dot{I} = \varepsilon k_0 V_0 \sin(k_0\theta - l_0\nu t + \phi)$$
$$\dot{\theta} = \omega(I) + \varepsilon \frac{\partial V_0}{\partial I} \cos(k_0\theta - l_0\nu t + \phi) \qquad (2.1.6)$$

where we have set

$$V_{k_0,l_0} = |V_{k_0,l_0}| \, e^{i\phi} = V_0 \, e^{i\phi}. \qquad (2.1.7)$$

It is also assumed that the value

$$\Delta I = I - I_0 \qquad (2.1.8)$$

is small, i.e., that the equations (2.1.6) are examined in the vicinity of the resonance value of the action I_0.

Here we list the approximations for further analysis:

(i) on the right-hand sides in equations (2.1.6) we may set $V_0 = V_0(I_0)$;
(ii) we may expand the frequency $\omega(I)$ as follows:

$$\omega(I) = \omega_0 + \omega' \Delta I$$

where

$$\omega_0 = \omega(I_0), \qquad \omega' = d\omega(I_0)/dI,$$

while ΔI has been defined in (2.1.8);
(iii) we may neglect the frequency term of the order of ε in the second equation (2.1.6).

Finally, we have reduced our system (2.1.6) to a simplified one:

$$\frac{d}{dt}(\Delta I) = \varepsilon k_0 V_0 \sin\psi$$
$$\frac{d}{dt}\psi = k_0 \omega' \Delta I \qquad (2.1.9)$$

where a new phase was introduced:

$$\psi = k_0\theta - l_0\nu t + \phi. \qquad (2.1.10)$$

The set of equations (2.1.9) can be presented in a Hamiltonian form:

$$\frac{d}{dt}(\Delta I) = -\frac{\partial \bar{H}}{\partial \psi}; \qquad \frac{d}{dt}\psi = \frac{\partial \bar{H}}{\partial(\Delta I)} \qquad (2.1.11)$$

where

$$\bar{H} = \tfrac{1}{2}k_0\omega'(\Delta I)^2 + \varepsilon k_0 V_0 \cos \psi. \qquad (2.1.12)$$

Expression (2.1.12) is an effective Hamiltonian describing the dynamics of a system in the neighbourhood of a resonance. Variables ΔI, ψ form a canonically conjugate pair.

A comparison of the expressions (2.1.12) and (1.4.1) shows that the Hamiltonian $\bar{H}$ describes the oscillations of a nonlinear pendulum (with the accuracy of the phase shift by π for $\omega' > 0$). It follows from (2.1.9) that

$$\ddot{\tilde{\psi}} + \Omega_0^2 \sin \tilde{\psi} = 0 \qquad (2.1.13)$$

where $\tilde{\psi} = \psi + \pi$ and

$$\Omega_0 = (\varepsilon k_0^2 V_0 |\omega'|)^{1/2}. \qquad (2.1.14)$$

The value Ω_0 is the frequency of phase oscillations. All the formulas obtained in Sect. 1.4 for the nonlinear pendulum can automatically be applied to the Hamiltonian (2.1.12) if we substitute $k_0\omega't$ for t in them.

On the phase plane (p, q) a phase curve defined by the action I_0 (see Fig. 2.1.1) corresponds to an exact resonance. Switching in equation (2.1.10) from polar angle θ to phase ψ takes us to a coordinate system rotating with the frequency of $l_0\nu$. Two samples of the phase portrait of the nonlinear resonance in the rotating coordinate system are shown in

Fig. 2.1.1   The nonlinear resonance at $k_0 = 1$, $l_0 = 1$ on the phase plane (p, q). The dotted line is the unperturbed trajectory with $I = I_0$. Thin lines are the phase oscillations. The thick line is the separatrix of the phase oscillations.

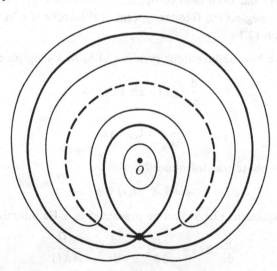

Fig. 2.1.1 and Fig. 2.1.2. It follows that in the case of a resonance of the order of l_0 on the phase plane, l_0 separatrix cells are formed with l_0 hyperbolic singular points and l_0 elliptic singular points. Thus, the nonlinear resonance transforms the topology of the phase portrait.

Let us determine the conditions under which the system is 'trapped' by a nonlinear resonance. We shall introduce the dimensionless parameter α characterizing the degree of nonlinearity of oscillations.

$$\alpha = \frac{I_0}{\omega_0} \left| \frac{d\omega(I_0)}{dI} \right| \equiv \frac{I_0}{\omega_0} |\omega'|. \tag{2.1.15}$$

The three approximations above are valid if the following conditions are satisfied

$$\varepsilon \ll \alpha \ll 1/\varepsilon. \tag{2.1.16}$$

The first inequality means that the nonlinearity must be fairly strong and should not tend to zero (Note 2.2).

From the Hamiltonian (2.1.12) and (2.1.15) we can estimate the amplitude of the phase oscillations expressed in action:

$$\frac{\max \Delta I}{I_0} \sim \left(\varepsilon \frac{V_0}{|\omega'|} \right)^{1/2} \frac{1}{I_0} \sim \left(\frac{\varepsilon}{\alpha} \right)^{1/2}, \tag{2.1.17}$$

while $V_0 \sim H_0 \sim \omega_0 I_0$. Similarly, with the help of the definition (2.1.14)

Fig. 2.1.2 Complicated version of the nonlinear resonance at $k_0 = 6$, $l_0 = 1$ (same notation as in Fig. 2.1.1).

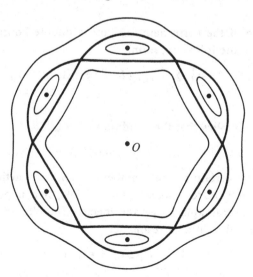

for the width of the resonance in frequency we get

$$\frac{\max \Delta\omega}{\omega_0} = \frac{\Omega_0}{\omega_0} = (\varepsilon\alpha)^{1/2}. \qquad (2.1.18)$$

Expression (2.1.17) explains the meaning of the first equation in (2.1.16), and expression (2.1.18) the meaning of the second equation, i.e., that the relative changes of action and frequency in the case of a nonlinear resonance must be small.

The impact of the non-resonance terms in the original expansion (2.1.3) will be discussed in the next section.

2.2 Internal nonlinear resonance

The approach towards resonance changes in a nonlinear system under the effect of an external periodic perturbation is fairly universal. It can be generalized relatively easily to the case of a resonance between two or more degrees of freedom.

Let us restrict ourselves to the case of $N = 2$. The Hamiltonian of a system with two degrees of freedom can be presented in the following form:

$$H = \sum_{i=1,2} H_{i0}(I_i) + \varepsilon V(I_1, I_2; \theta_1, \theta_2). \qquad (2.2.1)$$

The equations of motion are as follows:

$$\dot{I}_i = -\frac{\partial H}{\partial \theta_i}, \qquad \dot{\theta}_i = \frac{\partial H}{\partial I_i}. \qquad (2.2.2)$$

The expansion of the perturbation V into a double Fourier series can be expressed in the following form:

$$V(I_1, I_2; \theta_1, \theta_2) = \tfrac{1}{2} \sum_{n_1, n_2} V_{n_1, n_2}(I_1, I_2) \exp\{i(n_1\theta_1 - n_2\theta_2)\} + \text{c.c.} \qquad (2.2.3)$$

The resonance implies that the condition

$$n_1\omega_1(I_{10}) = n_2\omega_2(I_{20}) \qquad (2.2.4)$$

is satisfied for a certain pair of numbers (n_1, n_2) and actions (I_{10}, I_{20}).

If we retain only the resonance terms in (2.2.3), then, in perfect analogy to the previous paragraph, we get the effective Hamiltonian corresponding to the internal nonlinear resonance:

$$\bar{H} = \tfrac{1}{2} \sum_{i=1,2} \omega_i'(\Delta I_i)^2 + \varepsilon V_0 \cos \psi, \qquad (2.2.5)$$

where

$$\omega_i = \frac{dH_{i0}}{dI_i}; \qquad \omega_i' = \frac{d\omega_i(I_{i0})}{dI_i}; \qquad \Delta I_i = I_i - I_{i0} \qquad (2.2.6)$$

and phase ψ is defined by the following relations:

$$\psi = n_1 \theta_1 - n_2 \theta_2 + \phi$$

$$V_{n_1, n_2}(I_{10}, I_{20}) = |V_{n_1, n_2}(I_{10}, I_{20})| \, e^{i\phi} \equiv V_0 \, e^{i\phi}. \qquad (2.2.7)$$

The equations of motion follow directly from (2.2.5):

$$\dot{I}_1 = \frac{d(\Delta I_1)}{dt} = -\frac{\partial \bar{H}}{\partial \theta_1} = \varepsilon n_1 V_0 \sin \psi,$$

$$\dot{\theta}_1 = \frac{\partial \bar{H}}{\partial(\Delta I_1)} = \omega_1' \Delta I_1;$$

$$(2.2.8)$$

$$\dot{I}_2 = \frac{d(\Delta I_2)}{dt} = -\frac{\partial \bar{H}}{\partial \theta_2} = -\varepsilon n_2 V_0 \sin \psi,$$

$$\dot{\theta}_2 = \frac{\partial \bar{H}}{\partial(\Delta I_2)} = \omega_2' \Delta I_2.$$

Together with (2.2.7), the last two equations give us:

$$\dot{\psi} = n_1 \omega_1' \Delta I_1 - n_2 \omega_2' \Delta I_2. \qquad (2.2.9)$$

An integral of motion follows from the first two equations in (2.2.8)

$$n_1 \Delta I_1 + n_2 \Delta I_2 = \text{const} \equiv C. \qquad (2.2.10)$$

Together with equation (2.2.9), the latter makes the system (2.2.8) integrable. However, there is a simpler possibility.

Differentiating (2.2.9) with respect to time and making use of the first two equations in (2.2.8), we get:

$$\ddot{\tilde{\psi}} + \Omega_0^2 \sin \tilde{\psi} = 0, \qquad (2.2.11)$$

where

$$\Omega_0^2 = \varepsilon V_0 |n_1^2 \omega_1' + n_2^2 \omega_2'| \qquad (2.2.12)$$

and $\tilde{\psi} = \psi + \pi$, as before. Equation (2.2.11) is equivalent to the equation of the pendulum and the expression (2.2.12) defines the frequency of phase oscillations in the vicinity of an internal resonance.

2.3 The KAM theory

The above example of nonlinear resonance shows that even a very small perturbation may result in considerable changes in a system. Even when relatively small, these changes may nevertheless induce a significant qualitative restructuring of the system's properties. The major role in this belongs to nonlinearity. Since the frequency of oscillations is amplitude-dependent, the resonance cannot lead to an unlimited growth of the system's energy (or action), i.e., the resonance becomes mismatched. However, these results cannot guarantee the finite dynamics of the system for an infinite time. The question of eternal stability of a system was answered for the first time by the Kolmogorov–Arnold–Moser (KAM) theory with certain restrictions on a system's properties (Note 2.3).

Let us consider the Hamiltonian of a system in the following form:

$$H = H_0(I_1, \ldots, I_N) + \varepsilon V(I_1, \ldots, I_N; \theta_1, \ldots, \theta_N), \qquad (2.3.1)$$

where $\varepsilon \ll 1$ and V is the potential of a perturbation. In a certain sense, the form (2.3.1) is universal for the analysis of the effect of perturbations. The unperturbed part of the Hamiltonian H_0 depends on N actions which are independent first integrals of motion. Therefore, if $\varepsilon = 0$, there exist invariant N-dimensional tori, on whose surfaces all the system's trajectories lie. The frequencies of an unperturbed motion are defined by the following expression:

$$\omega_i = \frac{\partial H_0}{\partial I_i}, \qquad (i = 1, \ldots, N). \qquad (2.3.2)$$

Now, if we introduce a perturbation, i.e., assume that $\varepsilon \neq 0$, the question of a system's stability may be formulated in the following way: what will become of the invariant tori and indeed will they survive? We shall answer this question with the help of the Kolmogorov–Arnold theorem.

Theorem. If the unperturbed Hamiltonian system is non-degenerate, then, for a sufficiently small conservative perturbation, most of the non-resonance invariant tori will not disappear, but only undergo a slight deformation, so that in the phase space of the perturbed system there will also be invariant tori, everywhere densely filled with phase curves, which are wound around them conditionally-periodically, with the number of frequencies equal to the number of degrees of freedom. These invariant tori constitute a majority in the sense that the measure of complementation to their association is small, together with the perturbation.

Non-degeneracy of a system means the functional independence of its frequencies (2.3.2):

$$\det \left| \frac{\partial^2 H_0}{\partial I_i \, \partial I_j} \right| \neq 0. \tag{2.3.3}$$

We classify as non-resonance all the tori lying in a certain small region outside the immediate neighbourhood of a resonance, i.e.,

$$\left| \sum_j \omega_j m_j \right| > \text{const} |m|^{-(N+1)}$$

where const is a small value. Note that the definitions of non-degeneracy and non-resonance are given in terms of frequencies of the zero approximation (2.3.2).

The condition of sufficient smallness implies the existence of a certain boundary value ε_0, so that the theorem holds for $\varepsilon < \varepsilon_0$.

A similar assertion of existence of invariant tori, provided that a sufficiently large number of derivatives of the perturbation V exist, was proved by Moser (later it proved possible to reduce this number).

According to the KAM theory, the perturbation acts as follows. It destroys the tori located in the immediate vicinity of the resonance tori. One can estimate the size of this vicinity. For example, in the case of the first-order resonance between two degrees of freedom, discussed in Sect. 2.2, the width of the destruction region along the action axis is $< \varepsilon^{1/2}$. The subsequent properties of the regions of tori destruction also depend on the topology of the unperturbed tori in phase space. This leads to the case of $N = 2$ being different from that of $N > 2$. For $N = 2$, the tori divide phase space (see Sect. 1.5). Thus, the destroyed tori lie between the invariant tori. This results in the variation of the action along the trajectory always being small, since it has the characteristic length of the order of that of the destruction region. The trajectory is squeezed between the invariant tori. Its deviation from the unperturbed trajectory tends to zero as $\varepsilon \to 0$. Thus, the eternal and global stability for $N = 2$ and $\varepsilon < \varepsilon_0$ is proved.

If $N > 2$, the invariant tori no longer divide phase space. Thus, the destruction regions may join and permeate the entire phase space, just like a spider-web. This results in the appearance of a finite measure of trajectories, which may depart as far as desired from their non-perturbed counterparts. This phenomenon is known as Arnold diffusion [7] and will be discussed in Part II. Thus, for $N > 2$, the greater part of the tori are not destroyed. However, a finite, though small, measure of such initial conditions exists, which results in the system slowly departing as far as desired from its unperturbed trajectory.

The above results of the KAM theory can be easily applied to the already encountered case of the non-stationary system:

$$H = H_0(I_1, \ldots, I_N) + \varepsilon V(I_1, \ldots, I_N; \theta_1, \ldots, \theta_N; t). \quad (2.3.4)$$

Where V is a periodic function of time:

$$V(I, \theta; t + T) = V(I, \theta; t).$$

To accomplish this, we act by analogy with (1.1.3), introducing new variables (J, ϕ):

$$J = \text{const}; \qquad \phi = \nu t + \text{const}; \qquad \nu = 2\pi/T. \quad (2.3.5)$$

Now, instead of (2.3.4) we may consider a system with $(N+1)$ degrees of freedom:

$$H = H_0(I_1, \ldots, I_N) + \nu J + \varepsilon V(I_1, \ldots, I_N; \theta_1, \ldots, \theta_N; \phi). \quad (2.3.6)$$

The Hamiltonian H is a function of $N+1$ actions $(I_1, \ldots, I_N, J)$ and the same number of angles $(\theta_1, \ldots, \theta_N, \phi)$. The invariant tori are $(N+1)$-dimensional. The KAM theory ensures their stability for a sufficiently small ε, provided that the extended condition (2.3.3) is satisfied.

2.4 Local instability

The strongest form of instability, which is possible even in the case of finite motion, is local instability. It is due to this form of instability that dynamic stochasticity, or chaos, occurs.

Let $D(t)$ denote the distance between two points in phase space belonging to different trajectories at the moment of time t (Fig. 2.4.1). Then local instability can be defined as follows: there exists a direction along which the distance between the trajectories increases exponentially with time:

$$D(t) = D(0) \exp(h_0 t). \quad (2.4.1)$$

Fig. 2.4.1 The local instability of trajectories.

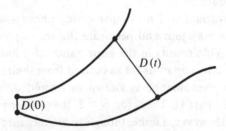

The increment of instability h_0 is a function of a point in phase space. The property reflected in equation (2.4.1) does not necessarily manifest itself, depending upon the initial conditions. However, local instability of a system implies that there exists a region of finite measure, so that if we arbitrarily choose one of its points as initial, a small perturbation will result in a significant divergence of the corresponding trajectories from the unperturbed ones. As a consequence, they get extremely tangled. The phase drop of regular shape soon acquires a complex shape (Fig. 2.4.2). The complexity of the drop's shape increases with time and may be described with the help of the mixing rate (see the next paragraph).

Let us introduce z, a vector defining a point in phase space of a system. Let $\hat{T}$ denote the mapping of a dynamic system performing finite motion:

$$z_{n+1} = \hat{T} z_n, \tag{2.4.2}$$

where n is discrete time. The Jacobian matrix $\hat{M}$ of a mapping $\hat{T}$ is defined by the following formula:

$$M_{ij,n} = \frac{\partial z_{i,n+1}}{\partial z_{j,n}}, \qquad (i, j = 1, \dots, P) \tag{2.4.3}$$

where P is the dimensionality of phase space. In the Hamiltonian systems, $P = 2N$. In addition, the phase volume of a Hamiltonian system is conserved and $|\hat{M}| = 1$ at any time t.

The eigenvalues λ of matrix $\hat{M}$ can be obtained from the equation

$$|\hat{M} - \lambda \hat{1}| = 0,$$

where $\hat{1}$ is the P-dimensional identity matrix. The values λ_k are, generally speaking, complex. They can be arranged in order of increasing absolute values:

$$|\lambda_1| \leqslant |\lambda_2| \leqslant \cdots \leqslant |\lambda_k| \leqslant \cdots \leqslant |\lambda_P|$$

Fig. 2.4.2 The spreading of a phase drop.

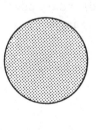
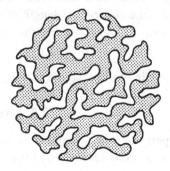

where λ_k is the last eigenvalue, for which $|\lambda_k| < 1$. This means that in phase space there exist (P_{-k}) directions, along which a perturbation of the state vector z stretches. An element of the phase volume stretches in the same directions.

The stretching index may be defined as follows. Let us, for simplicity's sake, assume that the stretching takes place in the direction of the ith eigenvector at the mth step of the mapping. Its eigenvalue is $|\lambda_{i,m}| > 1$. Let us further assume that λ_i does not depend on m. Then, after n steps of the mapping, an element of length d of the state vector will grow in that direction as

$$d_i(n) = |\lambda_i|^n d_i(0) = d_i(0) \exp(\sigma_i n) \qquad (2.4.4)$$

where the quantity

$$\sigma_i = \ln|\lambda_i| \qquad (2.4.5)$$

is the Lyapunov exponent.

If all the directions have Lyapunov exponents conserved in time, then local instability can be characterized by the expression:

$$h_0 = \sum_{i>k}^{P} \sigma_i = \sum_{i>k}^{P} \ln|\lambda_i|. \qquad (2.4.6)$$

Formulas (2.4.4)–(2.4.6) define a certain idealized situation of local instability. Lyapunov exponents in real Hamiltonian systems are, as a rule, functions of a point in space, the eigenvalues $\lambda_{i,n}$ of the Jacobian matrix $\hat{M}(n)$ depending also on time n. Moreover, in reality it is difficult to prove the validity of equation (2.4.4) (i.e., the existence of an asymptotic exponential law with $\sigma_i > 0$) not only for any n, but even on the average. Later, we shall learn what is responsible for these difficulties.

In the Hamiltonian case, $|\hat{M}| = 1$ and, therefore, $\lambda_1 \cdot \lambda_2 \cdots \lambda_{2N} = 1$. Specifically, for $N = 1$, we get:

$$\lambda_1 \lambda_2 = 1.$$

This suggests that in the case of instability, one dimension of phase space is unstable (the elements of length are stretched) and the other is stable (the elements of length are contracted). If, for example,

$$\lambda_1 = 1/\lambda_2 = \mathrm{const} > 1$$

then, according to (2.4.1) and (2.4.6), $h_0 = \lambda_1$.

2.5 Mixing

The property of local instability is immediately related to the property of mixing in phase space. Let $f(z)$ and $g(z)$ be two arbitrary integrable functions. The correlation function, or correlator, is the following:

$$\mathscr{R}(f, g; T) = \langle f(\hat{T}z)g(z)\rangle - \langle f(z)\rangle\langle g(z)\rangle \qquad (2.5.1)$$

where the angle brackets $\langle \cdots \rangle$ denote the average over phase space:

$$\langle f \rangle \equiv \int_\Gamma f(z)\, d\Gamma(z). \qquad (2.5.2)$$

In the case of ergodic motion, the equality of temporal and phase averages holds true:

$$\lim_{T \to \infty} \frac{1}{T} \int_t^{t+T} dt'\, f[z(t')] = \langle f \rangle \qquad (2.5.3)$$

Therefore, if the process is ergodic, averages $\langle f \rangle$ and $\langle g \rangle$ in (2.5.1) are time-independent.

The term 'mixing' refers to the decay (or decomposition) of correlators

$$\lim_{T \to \infty} \mathscr{R}(f, g; T) = 0. \qquad (2.5.4)$$

The law by which correlators decrease depends on the choice of functions f and g. The process $z(t)$, characterized by the property of mixing, will automatically be ergodic in the sense of definition (2.5.3). The reverse statement is, in the general case, not valid.

In the case of discrete time, the definition of the correlator has the form

$$\mathscr{R}_n(f, g) = \langle f(\hat{T}^n z), g(z)\rangle - \langle f(z)\rangle\langle g(z)\rangle \qquad (2.5.5)$$

and the mixing condition (2.5.4) is replaced by the following one:

$$\lim_{n \to \infty} \mathscr{R}_n(f, g) = 0. \qquad (2.5.6)$$

In this case the law by which correlator $\mathscr{R}_n$ decreases depends not only on the form of functions f and g, but also on the choice of a sequence of mapping times $t_0, t_1, \ldots, t_n$, as well.

The property of mixing implies that two kinds of fluid (the phase liquid and the region not yet filled with it) are steadily mixing (see Fig. 2.4.2). Henceforth, the exponential law of mixing will be implied, as a rule, i.e.,

$$\mathscr{R}(t) = \mathscr{R}_0 \exp(-t/\tau_c). \qquad (2.5.7)$$

The time τ_c is referred to as the mixing or correlator decay time.

If we omit some highly special and very difficult questions then, in a certain sense, the property of local instability automatically implies mixing. What is more, the following relation holds true:

$$h_0 \sim 1/\tau_c. \qquad (2.5.8)$$

Although intuitively the expression (2.5.8) seems quite natural, it can be proved only for a few 'straightforward' cases (Note 2.4).

The ability of trajectories of dynamic systems to have decomposing correlators or local instability means that they can be looked upon as a realization of random processes. This phenomenon was termed chaos, or stochasticity. Its characteristic feature is that although there are no random forces acting upon the system and no external random sources either, the system displays signs of the laws of stochastic motion. Moreover, different parameters of the system may determine different thresholds of the onset of local instability.

There is a certain concept of Kolmogorov-Sinai entropy (the KS entropy) which can be applied to systems possessing the property of mixing. Its definition was first introduced by Kolmogorov in [8] and was further refined by Sinai [9]. It is convenient to define the Kolmogorov-Sinai entropy by means of physical reasoning [1, 2]. Let us assume that a phase drop occupies volume $\Delta\Gamma$ in phase space. Its entropy then is:

$$S = \ln \Delta\Gamma. \tag{2.5.9}$$

If the initial phase volume of the drop is $\Delta\Gamma_0$, then by time t it occupies the volume $\Delta\Gamma(t)$. From Liouville's theorem on the conservation of the phase volume it follows that

$$\Delta\Gamma(t) = \Delta\Gamma_0$$

and the entropy (2.5.9) remains unchanged. Taking into consideration that the shape of the phase drop becomes more complex in the course of mixing, we introduce the notion of a coarsened phase volume

$$\overline{\Delta\Gamma(t)}.$$

The coarsened volume has small cavities free of phase liquid. As a result the coarsened phase volume increases with time, since Liouville's theorem is inapplicable.

Using the formula of local instability (2.4.1), we can estimate

$$\overline{\Delta\Gamma(t)} = \Delta\Gamma_0\, e^{ht} \tag{2.5.10}$$

where

$$h = \langle h_0 \rangle. \tag{2.5.11}$$

Now let us make use of equation (2.5.9) to obtain the entropy of the coarse grain phase drop.

$$\tilde{S} \equiv \ln \overline{\Delta\Gamma(t)} = \ln(\Delta\Gamma_0\, e^{ht}) = ht + \ln \Delta\Gamma_0. \tag{2.5.12}$$

We can set the value of $\Delta\Gamma_0$ equal to the small volume over which the averaging is done and where cavities can yet be neglected. The following expression

$$\lim_{\Delta\Gamma_0 \to 0} \lim_{t \to \infty} \frac{1}{t} \ln \overline{\Delta\Gamma(t)} = \lim_{\Delta\Gamma_0 \to 0} \lim_{t \to \infty} \frac{1}{t} (ht + \ln \Delta\Gamma_0) = h \qquad (2.5.13)$$

can be calculated with the help of (2.5.12). It defines the Kolmogorov-Sinai entropy. It is extremely important to pass to the limits in (2.5.13) strictly in the order prescribed.

Expression (2.5.13) defines the rate of changes of entropy $\tilde{S}$ and the coarsened volume. According to (2.5.11) and to (2.5.8), the following relation between mixing, local instability and the Kolmogorov-Sinai entropy is valid

$$h \sim h_0 \sim 1/\tau_c. \qquad (2.5.14)$$

Equation (2.5.14) plays a fundamental role in the analysis of conditions of the onset of chaos in dynamic systems.

Part II · Dynamic order and chaos

3
The stochastic layer

In the general case, Hamiltonial systems are carriers of chaos. This means that under minimal restrictions, the phase space of an arbitrary dynamic Hamiltonian system contains certain regions where motion is accompanied by mixing. These restrictions mostly concern the system's dimensionality. It must have no less than $\frac{3}{2}$ degrees of freedom, i.e., $N \geqslant \frac{3}{2}$. Let us henceforth always assume that time, being worth $\frac{1}{2}$ of one degree of freedom, is a cyclical variable. To put it otherwise, if the system's Hamiltonian is an explicit function of time, this dependency is a cyclical one.

The coexistence of regions of stable dynamics and regions of chaos in phase space, belongs to the most wonderful and striking discoveries. It enables us to analyse the onset of chaos and the appearance of the minimal region of chaos. Although much in this field still remains in question, nevertheless, it is evident that the seed of chaos is a stochastic layer which forms in the vicinity of destroyed separatrices. The stochastic layer was first described in [1], in a study on the stability and decomposition of magnetic surfaces, various estimations of its width being obtained later in [2]. The present chapter is dedicated to the systematic analysis of the stochastic layer and to several important applications of it (Note 3.1). The formation of a stochastic layer in the case of the perturbation of a nonlinear pendulum is typical. Therefore, we shall begin our discussion with this model.

3.1 The stochastic layer of a nonlinear pendulum:
mapping close to a separatrix

If the pendulum is affected by a non-stationary perturbation, its separatrix is destroyed and a stochastic layer forms in the neighbourhood. The process of its formation bears the characteristic features of the general case. Therefore, it is convenient to begin the discussion on the onset of chaos with this model.

Let us assume that the expression for the unperturbed term H_0 in the Hamiltonian

$$H = H_0(x, \dot{x}) + \varepsilon V(x, t) \tag{3.1.1}$$

relates to the nonlinear pendulum of unit mass (see equation (1.4.1)):

$$H_0 = \tfrac{1}{2}\dot{x}^2 - \omega_0^2 \cos x, \tag{3.1.2}$$

the perturbation V being a periodic function of time:

$$V(x, t + T) = V(x, t) \tag{3.1.3}$$

with the following period

$$T = 2\pi/\nu. \tag{3.1.4}$$

By means of definitions (1.3.1) and (1.3.3), we are able to change variables, switching from $(\dot{x}, x)$ to (I, θ), as follows:

$$I = \frac{1}{2\pi} \oint dq \, p(q, H_0); \qquad p(q, H_0) = [2(H_0 + \omega_0^2 \cos x)]^{1/2}. \tag{3.1.5}$$

The change of variables in equations (3.1.5) is done with the help of the unperturbed Hamiltonian H_0. All the necessary information has already been obtained in Sect. 1.4. Equations (1.4.10) or (1.4.11) define the pendulum's velocity $\dot{x}$ as a function of (I, θ, t) and provide the Fourier series expansion for it. The formula (1.4.8) substitutes $I(H_0)$ for I. Its reversion gives us the dependency $H_0 = H_0(I)$. Note that, while using formulas from Sect. 1.4, we should everywhere write H_0 instead of H.

With the help of the above substitution (3.1.5), the original Hamiltonian (3.1.1) can be presented in the following way:

$$H = H_0(I) + \varepsilon V(I, \theta, t). \tag{3.1.6}$$

Expression (3.1.6) is the general form of a system affected by a time-dependent perturbation. Special features of each particular problem show themselves in the form of functions H_0 and V.

Let us present an exact equation for the variation of energy H_0:

$$\dot{H}_0 = [H_0, H] = \varepsilon[H_0, V]$$

$$= -\varepsilon \frac{\partial H_0}{\partial \dot{x}} \frac{\partial V}{\partial x} = -\varepsilon \dot{x} \frac{\partial V}{\partial x}. \tag{3.1.7}$$

For reasons clarified below, it is convenient to use the variables $(\dot{x}, x)$ in equation (3.1.7).

Now, we are going to discuss motion in the vicinity of a separatrix. The main feature that will be exploited below, has to do with the behaviour of the velocity of unperturbed motion in the vicinity of a separatrix $\dot{x}$. Presented in Fig. 1.4.2, it looks like a sequence of pulses, each resembling a soliton's profile. The ratio of time interval between the pulses to their width is

$$N = \omega_0 / \omega(I) \gg 1 \qquad (3.1.8)$$

since $\omega(I) \to 0$ as I approaches its separatrix value

$$I_s = \frac{1}{\pi} 8\omega_0 E(\pi/2; 1) = \frac{1}{\pi} 8\omega_0 \qquad (3.1.9)$$

(see equation (1.4.8) for $\varkappa = 1$). The relation (3.1.8) shows that the system spends a considerable time ($\sim T = 2\pi/\omega$) in the vicinity of turning points where its velocity is close to zero, and quickly ($\sim T_0 = 2\pi/\omega_0$) traverses the greater part of a potential well.

The above properties of the pendulum's velocity $\dot{x}$, present on the right-hand side of equation (3.1.7), enable us to construct the mapping describing the dynamics in the vicinity of a separatrix. It is defined as the relation between the variables (I, θ) in the course of transform from one pulse of velocity to another.

For technical reasons, we make another change of variables, switching from (I, θ) to (E, t) where the following equation determines the transform

$$E = H_0(I). \qquad (3.1.10)$$

Since perturbation is a periodic function of time, we may introduce the phase of perturbation:

$$\varphi = \nu t + \text{const} \qquad (3.1.11)$$

which is an equivalent of time. Thus, the equations (3.1.10) and (3.1.11) define a new canonical pair (E, φ) which turns out to be very useful.

Let (E, φ) be values of the variables just before the velocity pulse (Fig. 3.1.1). Consequently, $(\bar{E}, \bar{\varphi})$ are values of these variables for half of the period of the pendulum's oscillations later. The desired mapping is now defined by the following relation:

$$(\bar{E}, \bar{\varphi}) = \hat{T}(E, \varphi) \qquad (3.1.12)$$

which may be presented in the following form:

$$\begin{aligned} \bar{E} &= E + \Delta E, \\ \bar{\varphi} &= \varphi + \pi\nu/\bar{\omega}, \end{aligned} \qquad (3.1.13)$$

where ΔE is the variation of energy $E = H_0$ due to the effect of velocity pulse, the frequencies being equal to $\omega = \omega(E)$ and $\bar{\omega} = \omega(\bar{E})$. In order to obtain these dependencies, we must substitute (3.1.10) in $\omega = \omega(I)$.

The value ΔE is easily expressed with the help of integration of (3.1.7):

$$\Delta E = -\varepsilon \int_{\Delta t} \mathrm{d}t\, \dot{x} \frac{\partial V}{\partial x} \qquad (3.1.14)$$

where the integral Δt is

$$\Delta t = \bar{T}/2 = \pi/\bar{\omega} \qquad (3.1.15)$$

in accordance with the second equation in (3.1.13). Its position on the time scale follows from Fig. 3.1.1.

Some of the singularities of mapping (3.1.13), also known as the separatrix mapping, may be understood without calculating ΔE. It is sufficient to notice that

$$|\Delta E| = |\bar{E} - E| \sim \varepsilon V \sim \varepsilon E. \qquad (3.1.16)$$

Therefore, with an accuracy up to small values of the higher order, we can use the unperturbed values of variables under the integral sign on the right-hand side of (3.1.14). Condition (3.1.16) also indicates that a change in energy cannot lead to a considerable instability, at least for a time interval $t \sim 1/\varepsilon$ (since $\Delta E \sim \varepsilon$). However, the second equation in (3.1.13) for the phase possesses the quite different property of an instability.

Let us consider the following expression

$$K = \left| \frac{\delta \bar{\varphi}}{\delta \varphi} - 1 \right| \qquad (3.1.17)$$

which determines the stretching of a small phase interval. If

$$K \gtrsim 1, \qquad (3.1.18)$$

Fig. 3.1.1   The choice of the sequence of time moments to construct a mapping in the vicinity of a separatrix.

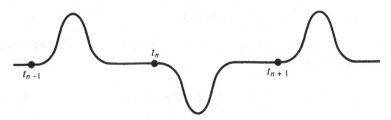

local instability in phase occurs. We are going to dwell on condition (3.1.18) in more detail later. Here let us assume only that it can serve as a good evaluation of the stochasticity region boundary. The main physical factor causing the onset of stochasticity (3.1.18) is as follows. A small energy change results in relatively small variations of frequency. Near the bottom of a potential well where the frequency is weakly energy(action)-dependent, its small changes also result in a small change in the phase over the period of oscillations. On the contrary, in the neighbourhood of a separatrix, where the period of oscillations $T = 2\pi/\omega$ tends to infinity, even small changes in frequency may result in considerable changes in phase. This is the very cause of local stochasticity.

3.2 The stochastic layer of a nonlinear pendulum: width of the layer

In this section we shall make a systematic evaluation of the width of a stochasticity region.

As follows from (3.1.13), ΔE must be calculated. To accomplish this, let us write out a typical presentation of the Hamiltonian of the unperturbed pendulum. Let us assume that

$$H = \tfrac{1}{2}\dot{x}^2 - \omega_0^2 \cos x + \varepsilon\omega_0^2 \cos(kx - \nu t) \qquad (3.2.1)$$

where ε is the dimensionless amplitude of perturbation while the pendulum's mass is equal to unity. The Hamiltonian (3.2.1) can be interpreted as the Hamiltonian of a particle moving in the field of two plane waves. Here expression (3.2.1) is presented in the frame of reference moving with the phase velocity of the first wave, while the phase velocity of the second wave in this frame of reference is ν/k. Parameter k designates the relation between two wave numbers.

The following equation of motion corresponds to the Hamiltonian (3.2.1):

$$\ddot{x} + \omega_0^2 \sin x = \varepsilon k\omega_0^2 \sin(kx - \nu t). \qquad (3.2.2)$$

Taking into consideration equation (3.2.2), the expression (3.1.7) assumes the following form:

$$\dot{E} = \varepsilon k\omega_0^2 \dot{x} \sin(kx - \nu t). \qquad (3.2.3)$$

The energy change according to (3.1.14) is obtained by way of integrating expression (3.2.3)

$$\Delta E = \varepsilon k\omega_0^2 \int_{\Delta t} dt\, \dot{x} \sin(kx - \nu t). \qquad (3.2.4)$$

Under the integral, one can use values of x and $\dot{x}$ for the motion on a separatrix, i.e., equations (1.4.4)-(1.4.6):

$$\dot{x} = \pm \frac{2\omega_0}{\cosh \omega_0(t - t_n)};$$

$$x = 4 \arctan \exp\{\pm\omega_0(t - t_n)\}$$

(3.2.5)

Here, it has been taken into account that the centre of the soliton is situated at a certain point t_n. Since $\dot{x}$ at $t \to \pm\infty$ decreases exponentially, it is possible to rewrite (3.2.4) in the following final form:

$$\Delta E = (-1)^n 2\varepsilon k \omega_0^2 \int_{-\infty}^{\infty} \frac{d\tau}{\cosh \tau} \sin\left(kx - \frac{\nu\tau}{\omega_0} - \varphi_n\right)$$

(3.2.6)

where the expression for phase φ_n is simple:

$$\varphi_n = \nu t_n$$

(3.2.7)

and corresponds to the phase coordinate of the soliton at a moment t_n in compliance with equation (3.1.11). In addition, the sign in (3.2.5) is assumed to be positive for even n, without affecting the generality. Expression (3.2.6) can be calculated by different methods, however, this being a purely technical question.

Let us present the value of ΔE for the important case of $k = 1$. To do this, let us note that the following relations result from equation (3.2.5):

$$\cos\frac{x}{2} = \frac{1}{\cosh \tau}; \qquad \sin\frac{x}{2} = \frac{\sinh \tau}{\cosh \tau}; \qquad \tau = \omega_0(t - t_n). \quad (3.2.8)$$

By making use of these relations, we get from (3.2.6)

$$\Delta E = (-1)^n 4\pi\varepsilon\nu^2 \frac{\exp(\pi\nu/2\omega_0)}{\sinh(\pi\nu/\omega_0)} \sin \varphi_n$$

(3.2.9)

Here note that expression (3.2.9) is valid in the vicinity of a separatrix both when $H_0 > H_s = \omega_0^2$ and when $H_0 < H_s = \omega_0^2$ (see equation (1.4.3)).

When frequencies of the perturbation are small ($\nu \ll \omega_0$), (3.2.9) leads to

$$\Delta E = (-1)^{n+1} 4\varepsilon\nu\omega_0 \sin \varphi_n, \qquad (\nu \ll \omega_0) \quad (3.2.10)$$

i.e., variations of energy are of the order of the amplitude of perturbation. On the contrary, when frequencies of the perturbation are large,

$$\Delta E = (-1)^{n+1} 8\pi\varepsilon\nu^2 \exp(-\pi\nu/2\omega_0) \sin \varphi_n. \quad (3.2.11)$$

For the sake of convenience, let us denote

$$\Delta E = \Delta E_s \sigma \sin \varphi \qquad (3.2.12)$$

where $\sigma = \pm 1$,

$$\Delta E_s = 4\pi\varepsilon\nu^2 \frac{\exp(\pi\nu/2\omega_0)}{\sinh(\pi\nu/\omega_0)}, \qquad (3.2.13)$$

and the subscript n at σ and φ is omitted for simplicity.

From the expression for the frequency of the unperturbed pendulum (1.4.9), in the vicinity of a separatrix we have, according to (1.4.13)

$$\omega(E) = \frac{\pi\omega_0}{\ln\left(\dfrac{32E_s}{|E - E_s|}\right)}, \qquad E_s = \omega_0^2. \qquad (3.2.14)$$

Now the separatrix mapping (3.1.13) assumes the following form:

$$\bar{E} = E + \Delta E_s \sigma \sin \varphi$$

$$\bar{\varphi} = \varphi + \frac{\nu}{\omega_0} \ln\left(\frac{32E_s}{|\bar{E} - E_s|}\right). \qquad (3.2.15)$$

The mapping (3.2.15) enables us to determine the parameter of local instability K and then turn to evaluation of the region of stochasticity. According to (3.1.17), we get from (3.2.15)

$$K = \frac{\nu}{\omega_0} \frac{\Delta E_s}{|E - E_s|} |\sin \varphi|$$

$$\approx \frac{4\pi\varepsilon\nu^3 \exp(\pi\nu/2\omega_0)}{\omega_0|E - E_s| \sinh(\pi\nu/\omega_0)} \qquad (3.2.16)$$

The condition $K \geqslant 1$ determines the so-called stochastic layer in the vicinity of the unperturbed separatrix (Fig. 3.2.1).

Fig. 3.2.1 The stochastic layer in the vicinity of a separatrix of a nonlinear pendulum.

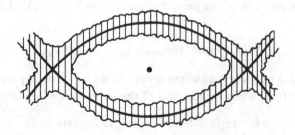

It is convenient now to turn our attention to the already mentioned limiting cases. In the case of small frequencies of perturbation ($\nu \ll \omega_0$), the value ΔE in the mapping (3.2.15) is determined by (3.2.10). Simple estimation shows the adiabatical law for the phase changing ($\ll 2\pi$) in (3.2.15) during almost all the period ($2\pi/\nu$) excluding a small interval of time in the vicinity of the moment of the separatrix crossing, and excluding the exponentially small area of energy

$$|E - E_s| < \omega_0^2 \exp(-\pi\omega_0/\nu). \qquad (3.2.17)$$

There is no rigorous theory of a stochastic layer width in this case, but estimations of different kinds give for the width

$$|E - E_s| \sim \max \Delta E = 4\varepsilon\nu\omega_0, \qquad (\nu \ll \omega_0). \qquad (3.2.18)$$

Since $E_s = \omega_0^2$, the relative width of a stochastic layer is of the order of $\varepsilon\nu/\omega_0$.

In the case of high frequencies of perturbation, $\nu \gg \omega_0$, from (3.2.16) it follows that

$$K = \frac{8\pi\varepsilon\nu^3}{\omega_0|E - E_s|} \exp(-\pi\nu/2\omega_0)|\sin\varphi|. \qquad (3.2.19)$$

Hence, for the border of stochasticity region it follows that

$$|E - E_s| \lesssim 8\pi\varepsilon \frac{\nu^3}{\omega_0} \exp(-\pi\nu/2\omega_0) \qquad (3.2.20)$$

and for the dimensionless width of the stochastic layer

$$\frac{\delta E}{E_s} = 8\pi\varepsilon \left(\frac{\nu}{\omega_0}\right)^3 \exp(-\pi\nu/2\omega_0) \qquad (3.2.20a)$$

where it has been taken into account that $E_s = \omega_0^2$. In this case the width of the stochastic layer proves to be exponentially small.

Finally, when we are trying to evaluate the width of the layer in the general case of (3.2.16), condition $K \gtrsim 1$ yields

$$|E - E_s| \lesssim \frac{4\pi\varepsilon\nu^3 \exp(\pi\nu/2\omega_0)}{\omega_0 \sinh(\pi\nu/\omega_0)} \qquad (3.2.21)$$

from whence the dimensionless width of the layer can be found:

$$\frac{\delta E}{E_s} = 4\pi\varepsilon \left(\frac{\nu}{\omega_0}\right)^3 \frac{\exp(\pi\nu/2\omega_0)}{\sinh(\pi\nu/\omega_0)}. \qquad (3.2.21a)$$

The above example of a stochastic layer introduced us to the phenomenon of minimal chaos, which consists of the following: in an arbitrary case, there always exists a stochastic region, no matter how small the parameter ε. The localization of the region of chaos is well defined. Its representative is the vicinity of separatrices. Later it will be shown that this statement is fairly universal.

Let us consider another example of a perturbation of the nonlinear pendulum:

$$H = \tfrac{1}{2}\dot{x}^2 - \omega_0^2 \cos x + \varepsilon \omega_0^2 x \sin \nu t. \qquad (3.2.22)$$

It corresponds to a dipole interaction between an oscillator and an external perturbation. For the model (3.2.22) we have:

$$V(x, t) = \omega_0^2 x \sin \nu t \qquad (3.2.23)$$

so that equation (3.1.14) for determining the energy change, assumes the following form:

$$\Delta E = -\varepsilon \int_{\Delta t} \mathrm{d}t\, \dot{x} \frac{\partial V}{\partial x}$$

$$= -\varepsilon \omega_0^2 \int_{\Delta t} \mathrm{d}t\, \dot{x} \sin \nu t. \qquad (3.2.24)$$

Now we have to repeat the procedure we used to calculate expression (3.2.4). The substitution of (3.2.5) into (3.2.24) and the calculation of the integral yield

$$\Delta E = \frac{2\pi\varepsilon\omega_0^2}{\cosh(\pi\nu/2\omega_0)}\, \sigma \sin \varphi. \qquad (3.2.25)$$

Therefore, ΔE_s in the separatrix mapping (3.2.15) is

$$\Delta E_s = \frac{2\pi\varepsilon\omega_0^2}{\cosh(\pi\nu/2\omega_0)}. \qquad (3.2.26)$$

This determines the border of the stochastic layer.

The formula obtained has the same characteristics as expression (3.2.21), and, therefore, there is no need for further discussion.

$$|E - E_s| \leqslant \frac{\nu}{\omega_0}\Delta E_s = \frac{2\pi\nu\varepsilon\omega_0}{\cosh(\pi\nu/2\omega_0)}.$$

3.3 Weak interaction of resonances

The results obtained in the previous paragraph allow a very simple presentation of the following example, describing the interaction of two resonances, lying far from each other [8]. The equation of motion for this system has the form

$$\ddot{x} + 2\omega_0^2 \cos(\nu t/2) \sin x = 0. \qquad (3.3.1)$$

It corresponds to the nonlinear pendulum with an oscillating frequency. The following inequality is assumed for the value of ν

$$\nu \gg \omega_0 \qquad (3.3.2)$$

i.e., the frequency of modulation $\nu/2$ is high enough compared to the frequency of small oscillations of the pendulum. If we present (3.3.1) in the following form:

$$\ddot{x} + \omega_0^2[\sin(x + \nu t/2) + \sin(x - \nu t/2)] = 0, \qquad (3.3.3)$$

it can be easily noticed that the system (3.3.3) is equivalent to the equation of motion (3.2.2), after the following substitution:

$$\tilde{x}_{\pm} = x \pm \nu t/2. \qquad (3.3.4)$$

The following Hamiltonian corresponds to the equations of motion (3.3.1) or (3.3.2):

$$\begin{aligned} H &= \tfrac{1}{2}\dot{x}^2 - 2\omega_0^2 \cos(\nu t/2) \cos x \\ &= \tfrac{1}{2}\dot{x}^2 - \omega_0^2[\cos(x - \nu t/2) + \cos(x + \nu t/2)]. \end{aligned} \qquad (3.3.5)$$

If we present phase portraits on a uniform scale and on the same plane, we get the already familiar pattern constituted by separatrix cells for the pendulum (Fig. 3.3.1). The only difference is that now we have two separatrix chains which do not intersect due to inequality (3.3.2). Indeed, the width of each separatrix is equal to ω_0 and the distance between them is ν. Inequality (3.3.2) also implies the absence of first-order resonance in the system $\nu \approx \omega_0$. It was this circumstance that always allowed the application of various approximate methods based on averaging over fast modulations. The theory of the stochastic layer contains new possibilities which enables us to answer an extremely subtle question: what information on the behaviour of systems (3.3.1) and (3.3.3) is lost in the course of the averaging procedure?

To answer this question, let us compare the equations of motion (3.2.2) and (3.2.3). They coincide after the following substitutions

$$x \to \tilde{x}_+, \qquad \varepsilon = -1.$$

Now let us note that equation (3.2.2) holds not only for small ε but also for $\varepsilon \sim 1$, due to an exponentially small factor in the energy change ΔE in the case of $\nu \gg \omega_0$ (see equations (3.2.12) and (3.2.13)). This enables us to apply expression (3.2.20). Thus, two interacting and non-overlapping resonances with the Hamiltonian (3.3.5) are dressed with stochastic layers in the vicinity of their destroyed separatrices (3.3.1). The width of the layers is exponentially small and equals

$$\delta E = 8\pi \frac{\nu^3}{\omega_0} \exp(-\pi\nu/2\omega_0). \tag{3.3.6}$$

The above result leads to a still more important conclusion. Let us consider the Hamiltonian of a particle moving in a field of multiple plane waves:

$$H = \tfrac{1}{2}\dot{x}^2 + \sum_k V_k \cos(kx - \omega_k t). \tag{3.3.7}$$

This problem has many applications [8–11]. Let us make the following assumptions concerning the form of potential in (3.3.7):

(i) amplitudes of V_k satisfy the following condition:

$$\sum_k V_k^2 < \infty \tag{3.3.8}$$

(ii) there is a considerable mismatch between two adjacent frequencies $\Delta\omega_k = |\omega_{k_1} - \omega_{k_2}|$:

$$\Delta\omega_k \gg |V_k|^{1/2}. \tag{3.3.9}$$

Fig. 3.3.1   Two non-overlapping resonances.

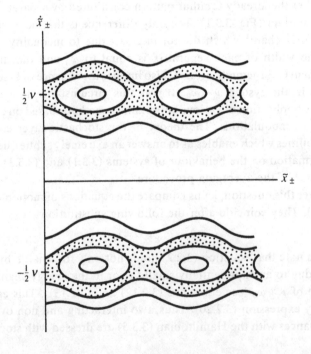

Condition (3.3.9) must be satisfied for all k. It implies the absence of resonance overlapping and is equivalent to inequality (3.3.2). When (3.3.38) and (3.3.9) hold true, then, on the phase plane, a particle moving in the vicinity of a plane wave is only slightly affected by other waves. In any case, the influence of other waves can be estimated by means of the usual averaging techniques. The only exception is the exponentially narrow region in the vicinity of separatrices of waves, where the influence of the neighbouring waves is considerable and the motion is stochastic (Note 3.2).

3.4 The standard mapping

This mapping, also referred to as the Chirikov-Taylor mapping, emerges in many physical problems [11, 6, 10]. At the same time, it offers one of the simplest models of the onset of chaos, retaining all the most typical and complex features of the problem.

The Hamiltonian of a system that produces the standard mapping, has the following form:

$$H = \tfrac{1}{2}I^2 - K \cos \theta \sum_{n=-\infty}^{\infty} \delta\left(\frac{t}{T} - n\right). \qquad (3.4.1)$$

It defines the system with an unperturbed Hamiltonian

$$H_0 = \tfrac{1}{2}I^2 \qquad (3.4.2)$$

that is affected by a periodic sequence of kicks (δ-pulses) with the period

$$T = 2\pi/\nu. \qquad (3.4.3)$$

Expression (3.4.2) corresponds either to a particle's free motion $I = p$, $\theta = x$ or to a free rotator, when the variable θ is cyclical, i.e., $\theta \in (0, 2\pi)$. Let us consider the latter case. By taking into account the following identity

$$\sum_{n=-\infty}^{\infty} \delta\left(\frac{t}{T} - n\right) = \sum_{n=-\infty}^{\infty} \cos\left(2\pi n \frac{t}{T}\right) \qquad (3.4.4)$$

the expression (3.4.1) can be rewritten in the following form:

$$H = \tfrac{1}{2}I^2 - K \cos \theta \sum_{m=-\infty}^{\infty} \cos m\nu t. \qquad (3.4.5)$$

The relation between (3.3.7) and (3.4.5) can be easily established. To accomplish this, it is sufficient to substitute the following values in equation (3.3.7):

$$k \equiv k_0 = 1; \qquad V_k \equiv V_0 = K; \qquad \omega_k = k\nu \qquad (3.4.6)$$

assuming k to be an integer and extending the summation over k to the range of $(-\infty, \infty)$. Thus, the Hamiltonian (3.4.1) corresponds, specifically, to a particle's motion in a periodic wave packet with an infinite number of harmonics of equal amplitude.

The equations of motion from (3.4.1) have the form:

$$\dot{I} = -K \sin \theta \sum_{n=-\infty}^{\infty} \delta\left(\frac{t}{T} - n\right)$$

$$\dot{\theta} = I \qquad\qquad\qquad (3.4.7)$$

between two δ-functions

$$I = \text{const}; \qquad \theta = It + \text{const}.$$

At each step, or kick, represented by a δ-function, the variable θ remains continuous and action I changes by the value $-K \sin \theta$. Assuming (I, θ) to be the values of variables just before an nth kick, $(\bar{I}, \bar{\theta})$ being their values before the next - $(n+1)$th - kick, then from (3.4.7) the mapping,

$$\bar{I} = I - K \sin \theta$$

$$\bar{\theta} = \theta + \bar{I}, \quad (\text{mod } 2\pi) \qquad\qquad (3.4.8)$$

follows, which is equivalent to the equations of motion (3.4.7).

We shall be mainly interested in the case of small perturbations:

$$K \ll 1. \qquad\qquad\qquad (3.4.9)$$

However, even with this simplification provided, equations (3.4.8) still appear extremely sophisticated. We can make sure of this immediately, prior to presenting any results of their analysis. In order to do this, let us turn to Fig. 3.4.1, which demonstrates the phase portrait of the system (3.4.8) for $K = 0.9$. It is comprised of the following elements. The most conspicuous in a separatrix cell with the saddle at $\theta = \pm\pi$ (the phase portrait should be thought of as a torus with the right and left, and upper and lower sides of the square pasted together in pairs). The separatrix has been destroyed, and a stochastic layer formed in its place. Inside this main stochastic layer lies a family of invariant curves enclosed one inside the other and embracing the point $(0, 0)$. Outside the main stochastic layer, there are necklaces of separatrix cells with considerably thinner stochastic layers. They correspond to nonlinear resonances of different orders and enclose the torus normal to the I axis. Between the resonances lie invariant curves enclosing the torus. Thus, the phase portrait resembles a sandwich composed of an infinite number of invariant curves alternating with stochastic layers.

An important feature of the phase portrait is that, provided the condition of smallness of K (3.4.9) is satisfied, stochastic layers do not merge. This follows directly from the KAM theory for the number of degrees of freedom $N \leqslant 2$. In our case, $N = \frac{3}{2}$ and invariant tori divide the phase space. The results obtained in the preceding sections of this chapter, offer an explanation of the form of the phase portrait presented above.

First, let us note that for small K, finite-difference equations (3.4.8) can be replaced by differential equations

$$\frac{\mathrm{d}I}{\mathrm{d}n} = -K' \sin \theta; \qquad \varphi \frac{\mathrm{d}\theta}{\mathrm{d}n} = I$$

which correspond to the equation of a pendulum:

$$\frac{\mathrm{d}^2\theta}{\mathrm{d}n^2} + K \sin \theta = 0. \tag{3.4.10}$$

Retaining in the sum in equation (3.4.5) only the terms with $m = 0, \pm 1$,

Fig. 3.4.1 The phase portrait for a standard mapping (numerical results) at $K = 0.9$: horizontal axis $\theta \in (-\pi, \pi)$; vertical axis $I \in (\pi, -\pi)$.

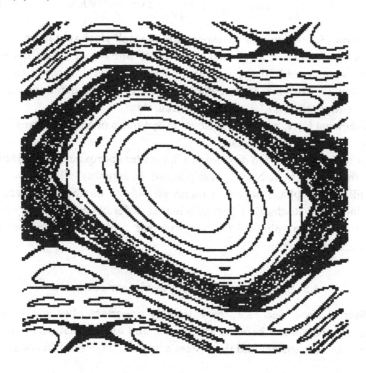

we get the following:

$$H \approx \tfrac{1}{2}I^2 - K \cos \theta - 2K \cos \nu t \cos \theta$$
$$= \tfrac{1}{2}I^2 - K \cos \theta - 2K \cos 2\pi n \cdot \cos \theta \qquad (3.4.11)$$

where n stands for time (see (3.4.10)). Equation (3.4.10) describes the pendulum's oscillations with the frequency

$$\omega_0 = \sqrt{K}. \qquad (3.4.12)$$

The destroyed separatrix of these oscillations forms the main stochastic layer in Fig. 3.4.1. The perturbation is given by the third term in (3.4.11). It has the same amplitude as the main term in the Hamiltonian H, however its frequency

$$\bar{\nu} = 2\pi \qquad (3.4.13)$$

is large if compared to ω_0, since $K \ll 1$. This enables us to make use of the formula (3.2.20) for the width of the stochastic layer. After substituting

$$\varepsilon = 2K, \qquad \omega_0 = \sqrt{K}, \qquad E_s = \omega_0^2 = K, \qquad \nu = \bar{\nu} = 2\pi$$

we get the following:

$$\frac{\delta E}{E_s} = \max \left| \frac{E - E_s}{E_s} \right| = \frac{(4\pi)^4}{2\sqrt{K}} \exp(-\pi^2/\sqrt{K}). \qquad (3.4.14)$$

Equation (3.4.14) defines an exponentially small stochastic layer, which forms in the case of any small value of K. Should we retain the next terms with $m = \pm 2$ in the sum in (3.4.5), the resulting corrections to equation (3.4.14) are exponentially small. Thus, the definition of the exponential term in (3.4.14) is very accurate.

Somewhat later, we shall see how the dynamics of a rotator transforms as the parameter K increases.

We have hitherto been discussing a first-order approximation, reflecting only the rough fabric of the phase portrait. In actuality, all windows and islands of the phase portrait contain an infinite number of separatrix necklaces and stochastic layers of a higher order. As a whole, they form an extremely complicated fractal ornament, which no one as yet has been able to describe.

3.5 Stochastic layer of a nonlinear resonance

In Sect. 2.1 and Sect. 2.2, the extremely important phenomenon of a nonlinear resonance between the system and an external perturbation, or between the system and the system's internal degrees of freedom, was considered. Among the simplifications used in the course of its analysis, there was one we are now going to discuss – the neglecting of non-resonant terms. In almost every field of modern physics, this approximation has become a common routine. In actual fact, it is by taking into account non-resonant terms that we arrive at a small but qualitatively new result – the formation of a stochastic layer in place of a separatrix of a nonlinear resonance. Because of the above results, we are already able to write down the width of this layer.

With this aim in view, note that the non-resonant term has been eliminated from the equations of motion (2.1.6). If we take this fact into account, the equation for the action assumes the following form:

$$\dot{I} = \varepsilon k_0 [\, V_0 \sin(k_0 \theta - l_0 \nu t + \varphi) + V_1 \sin(k_0 \theta + l_0 \nu t + \varphi)], \quad (3.5.1)$$

where

$$V_{k_0, -l_0} = |V_{k_0, -l_0}| \exp(i\varphi_1) \equiv V_1 \exp(i\varphi_1).$$

In equation (3.5.1) k_0 and l_0 are positive, therefore the second term is non-resonant. Expression (2.1.12) for the Hamiltonian of nonlinear resonance is transformed into the following one

$$H = \tfrac{1}{2} k_0 \omega'(\Delta I)^2 + \varepsilon k_0 V_0 \cos \psi$$
$$+ \varepsilon k_0 V_1 \cos(\psi + 2l_0 \nu t + \varphi_1 - \varphi). \quad (3.5.2)$$

Making use of the Hamiltonian equations of motion (2.1.11), we get from (3.5.2):

$$\frac{d \, \Delta I}{dt} = -\frac{\partial H}{\partial \psi} = \varepsilon k_0 V_0 \sin \psi + \varepsilon k_0 V_1 \sin(\psi + 2l_0 \nu t + \varphi_1 - \varphi)$$

$$\frac{d\psi}{dt} = \frac{\partial H}{\partial(\Delta I)} = k_0 \omega' \, \Delta I.$$

Hence it follows

$$\frac{d^2 \psi}{dt^2} + \Omega^2 \sin \psi = -\Omega_1^2 \sin(\psi + 2l_0 \nu t + \varphi_1 - \varphi) \quad (3.5.3)$$

where Ω_0 is the frequency of phase oscillations (see 2.1.14):

$$\Omega_0^2 = \varepsilon k_0^2 V_0 |\omega'| \quad (3.5.4)$$

and similarly,

$$\Omega_1^2 = \varepsilon k_0^2 V_1 |\omega'|.$$

Further, for simplicity of notation, we set $k_0 = 1$. The resonance condition is expressed by the following equality

$$\omega = l_0 \nu. \tag{3.5.5}$$

Therefore, $\nu \sim \omega$ and, consequently,

$$\nu \gg \Omega_0. \tag{3.5.6}$$

Inequality (3.5.6) is obvious since the frequency of the non-resonant term is of the same order as the frequency of unperturbed oscillations and the frequency of phase oscillations Ω_0 is proportional to the square root of perturbation and is therefore small. If we take advantage of the relation between the frequencies (2.1.18), (3.5.5) yields

$$\nu = \frac{1}{l_0}\,\omega = \frac{\Omega_0}{l_0(\varepsilon\alpha)^{1/2}} \tag{3.5.7}$$

where α is the parameter of nonlinearity of the problem.

Now let us compare the two equations of motion (3.2.2) and (3.5.3). They are identical, provided the following substitution is performed in (3.2.2):

$$\omega_0 \to \Omega_0; \qquad k \to 1; \qquad \varepsilon \to \Omega_1^2/\Omega_0^2; \qquad \nu \to 2l_0\nu.$$

Taking into account inequality (3.5.6) and making the substitution in equation (3.2.20) we get for the dimensionless width of the stochastic layer

$$\frac{\delta E}{E_s} \equiv \max \frac{|E - E_s|}{E_s} = 8\pi \left(\frac{\Omega_1}{\Omega_0}\right)^2 \left(\frac{2l_0\nu}{\Omega_0}\right) \exp(-\pi l_0\nu/2\Omega_0)$$

$$= \frac{64\pi}{(\varepsilon\alpha)^{3/2}} \left|\frac{V_1}{V_0}\right|^2 \exp\{-\pi/(\varepsilon\alpha)^{1/2}\}. \tag{3.5.8}$$

Specifically, for $|V_1| = |V_0|$, we simply have

$$\frac{\delta E}{E_s} = \frac{64\pi}{(\alpha\varepsilon)^{3/2}} \exp\{-\pi/(\varepsilon\alpha)^{1/2}\}. \tag{3.5.9}$$

This is an exponentially small expression of the same type as equation (3.4.14). It constitutes a universal result: each nonlinear resonance (an internal one, as well) is surrounded by a stochastic layer with thickness given by (3.5.8) or (3.5.9). Later, it will be shown in which way this important result affects some of the universal properties of dynamic systems.

3.6 Non-trivial effects of discretization

The above examples of the formation of a stochastic layer offer a good opportunity to discuss one of the most important questions posed by the rapid development of computational methods in various problems of the natural sciences. The basis of numerical analysis is to form finite-difference algorithms and, consequently, we are forced to transfer from differential equations to equations in finite differences. For example, the equation of motion for the pendulum:

$$\ddot{x} + \omega_0^2 \sin x = 0 \tag{3.6.1}$$

in the simplest case can be substituted by the following equation:

$$x_{n+1} - 2x_n + x_{n-1} + \omega_0^2 (\Delta t)^2 \sin x_n = 0. \tag{3.6.2}$$

Here Δt is the duration of the time interval that serves as an elementary step in computations and

$$x_n = x(n \, \Delta t).$$

To improve the accuracy of computations, very small values of Δt are chosen, so that the following inequality holds true:

$$K = \omega_0^2 (\Delta t)^2 \ll 1. \tag{3.6.3}$$

To what extent is one able to control the errors arising in the transfer from problem (3.6.1) to problem (3.6.2)? The question has a long history. Already the first numerical simulations of nonlinear physical problems have brought about the following discussion: what properties are lost and what do we gain by introducing discretization in nonlinear problems [12]. To a large measure, it became possible to answer these questions only after it became clear that chaos is possible in dynamic systems. The following example based on equation (3.6.1) enables us to clear up one universal and non-trivial effect of discretization.

Let us denote

$$p_n = \frac{1}{\Delta t} (x_n - x_{n-1}). \tag{3.6.4}$$

Now we rewrite (3.6.2) and (3.6.4) as a system

$$\begin{aligned} p_{n+1} &= p_n - \omega_0^2 \, \Delta t \sin x_n \\ x_{n+1} &= x_n + \Delta t \, p_{n+1}. \end{aligned} \tag{3.6.5}$$

Mapping (3.6.5) coincides with the standard mapping (3.4.8) if we replace $\Delta t \, p \to I$, $x \to \theta$ and use the notation (3.6.3). Therefore, we are already able to formulate a number of conclusions. The phase space of the system (3.6.1) consists only of invariant curves. On the contrary, the phase space of the system (3.6.5) contains an infinite number of stochastic layers, in the case of arbitrarily small values of K.

The transition to discrete equations is equivalent to the introduction of an external periodic force. Notice that mapping (3.6.5) is generated by the Hamiltonian (3.4.5), where:

$$I = p \, \Delta t, \qquad \theta = x$$

i.e.,

$$H = \tfrac{1}{2}p^2 - \omega_0^2 \cos x \sum_{m=-\infty}^{\infty} \cos m\nu t \qquad (3.6.6)$$

where

$$\nu = \frac{2\pi}{\Delta t}.$$

Taking into consideration inequality (3.6.3), let us confine ourselves in (3.6.6) only to the terms with $m = 0, \pm 1$. This gives us:

$$H = H_0 + V_{\text{discr}} \qquad (3.6.7)$$

where

$$H_0 = \tfrac{1}{2}p^2 - \omega_0^2 \cos x;$$
$$V_{\text{discr}} = -2\omega_0^2 \cos x \cos \nu t. \qquad (3.6.8)$$

These formulas show that H_0 is the Hamiltonian of the original equation (3.6.1) and V_{discr} is the potential of the perturbation caused by discretization. It results in the formation of a stochastic layer (see equation (3.4.14)) with the following relative thickness

$$\frac{\delta E}{E_s} = \frac{\delta E}{\omega_0^2} = \frac{(4\pi)^4}{2\omega_0 \, \Delta t} \exp(-\pi^2/\omega_0 \, \Delta t). \qquad (3.6.9)$$

This simple expression reflects a new qualitative aspect of discretization. The potential of V_{discr} is a high-frequency one ($\nu \gg \omega_0$), but its amplitude is of the same order as the amplitude of the unperturbed potential. All corrections due to V_{discr} are as small as corrections of high-frequency perturbations. This statement, however, holds only for perturbations of the invariant curves situated far from separatrices. In the vicinity of a separatrix the influence of V_{discr} results in qualitatively new dynamics, namely, in stochastic instability. In the case of a high-order dimensionality of the problem under consideration, the entire phase space is covered with a web of stochastic layers formed in place of destroyed separatrices. The existence of such a web, as we shall see later, may result in a considerable difference between the discrete problem and the continuous one.

3.7 Chaotic spinning of satellites

Phenomena which are, to a certain extent, related to chaos emerge most unexpectedly in domains of physics which have hitherto been looked upon as characterized by a high predictability of results.

In the series of works [13–15] Wisdom noticed that, as they travel along non-circular orbits, asymmetrical satellites may be executing stochastic spinning. The asymmetrical shape of satellites results in a spin-orbital coupling. One can name several natural objects for which these results may be of real interest, for example, Saturn's satellite Hyperion and Mars's satellite Phobos. Setting aside the astrophysical problems concerning Hyperion and Phobos, we shall consider only the dynamical aspect of the problem, which is of a certain interest for artificial satellites, as well.

For simplicity, we assume that the rotation axis of a satellite (spin axis) is perpendicular to the orbital plane (Fig. 3.7.1). The satellite itself is assumed to have the form of a three-axis ellipsoid with the main moments of inertia $A < B < C$ (C being the moment relative to the spin

Fig. 3.7.1 The geometry of spin-orbital coupling of a satellite.

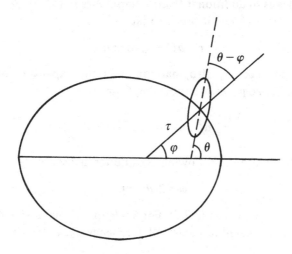

axis). The degree of asymmetry of the satellite can be characterized by the following parameter:

$$\alpha_0^2 = \frac{3}{C}(A - B). \tag{3.7.1}$$

The satellite's orbit is assumed to be of an elliptic form and in the polar coordinates (r, φ) has the following form:

$$r = \frac{a(1 - e_0^2)}{1 + e_0 \cos \varphi} \tag{3.7.2}$$

where a is the major semiaxis of the orbit and e_0 is its eccentricity. The frequency of orbital revolution is

$$\omega_0 = \frac{(\gamma M)^{1/2}}{a^{3/2}}$$

where γ is the gravitational constant and M is the planet's mass. Introducing the dimensionless time $\tau = \omega_0 t$, the equation for a satellite's spin motion is reduced to

$$\frac{d^2\theta}{d\tau^2} + \tfrac{1}{2}\alpha_0^2 \left(\frac{a}{r}\right)^3 \sin 2(\theta - \varphi) = 0. \tag{3.7.3}$$

In this equation, r and φ are functions of time. In the case of a circular orbit, we should simply have

$$\varphi = \omega_0 t = \tau; \qquad r = a. \tag{3.7.4}$$

It is the ellipticity of the orbit that makes r and φ time-dependent in a way which leads to additional Fourier harmonics in (3.7.3). Specifically, when $e_0 \ll 1$, from (3.7.2) it follows that

$$r \approx a(1 - e_0 \cos \tau).$$

Taking into account, also, harmonics in the expansion of φ [9], equation (3.7.3) acquires the following form

$$\frac{d^2\psi}{d\tau^2} + \alpha_0^2 \sin \psi = -\varepsilon \alpha_0^2 \sin(\psi - \tau), \tag{3.7.5}$$

where $\varepsilon = 7e_0/2$ and the following substitution is used

$$\psi = 2(\theta - \tau). \tag{3.7.6}$$

Let us analyse equation (3.7.5). For $\varepsilon = 0$, the left-hand side describes oscillations of the satellite's axis with frequency α_0. In the case of small asymmetry of the satellite's shape ($\alpha_0 \ll 1$), as for Mercury ($\alpha_0 = 0.017$), from (3.2.6) it follows that

$$\frac{d\psi}{d\tau} = \alpha_0 = 2\left(\frac{d\theta}{d\tau} - 1\right).$$

This relation signifies that the frequency of perturbation in (3.7.5) is equal to 1, i.e., the perturbation is of high frequency ($\alpha_0 \ll 1$). Such perturbation generates a stochastic layer in the vicinity of a separatrix with the width defined by equation (3.2.20a), where we have to assume

$$\varepsilon \to 7e_0/2; \qquad \nu \to 1; \qquad \omega_0 \to \alpha_0.$$

Hence

$$\frac{\delta E}{E_s} = \frac{28e_0}{\alpha_0^3} \exp(-\pi/2\alpha_0). \qquad (3.7.7)$$

For Mercury, $e_0 = 0.206$ and the region of chaotic dynamics has the order of magnitude 10^{-34}. This is by far too small a value. However, for Hyperion, $e_0 = 0.1$ and $\alpha_0 = 0.89$. The stochastic layer is of the order of unity and equation (3.7.7) does not work. Numerical analysis to define the region of stochasticity of the initial equation (3.7.3) for Hyperion has shown that we have the overlapping of resonances and formation of a large stochastic sea [13] at $e_0 = 0.1$ and $\alpha_0 = 0.89$. We are going to dwell upon the stochastic sea later.

Thus, there is room for chaotic behaviour in describing celestial objects in the solar system due to the spin-orbital coupling. General considerations presented here are also applicable to artificial satellites.

4
Stochastic layer to stochastic sea transition

The picture of the onset of chaos, as we see it today, is so extremely complex that one is advised to assume a step-by-step approach in its comprehension, temporarily omitting certain questions from consideration. In the preceding chapter, nothing was said of the border between the stochastic layer and the region of invariant curves. For that reason, in calculating the width of the stochastic layer, we adopted the approximate inequality $K \geqslant 1$ as the border of chaos. Of course, the question of the conditions of the onset of chaos emerged as soon as the first studies on the analysis of real physical systems appeared. The word 'real' here means 'typical' for many physical problems, since there are strict mathematical criteria of chaos which can be illustrated by not-too-abstract models (Note 4.1). The main feature shared by these 'chosen' cases is the absence in phase space of sandwich-like or hole-riddled structures in which regions of chaos alternate with regions of stability. Examples of the first and second type are presented in Figs. 3.4.1 and 4.0.1, respectively: the 'stochastic sea' (the shaded area) is made up of points of the mapping belonging to a single trajectory, while the light regions are islands that cannot be reached by a trajectory from the sea region, and vice versa.

The coexistence of stability regions and regions of chaos in phase space presents severe problems in the study of dynamical chaos (Note 4.2). This difficulty comes up, for example, in the structure of the stochastic sea. Actually, every island, together with its close vicinity, in Fig. 4.0.1, has a structure no less complex than that of the entire region depicted in Fig. 4.0.1 (see Fig. 4.0.2). The interior of an island is complex and possesses very narrow sandwich-like regions, etc. Patterns of this type are called fractals (Note 4.3). The hierarchic complexity of the phase portrait also applies to the method of its analysis. For example, if we

Fig. 4.0.1 The phase portrait for a standard mapping at $K = 1.2$; horizontal axis $\theta \in (-\pi, \pi)$; vertical axis $I \in (\pi, -\pi)$.

Fig. 4.0.2 Two consecutive enlargements of the region in phase space for a standard mapping at $K = 1.2$: (a) the region inside the square in Fig. 4.0.1; (b) the region inside the square in (a).

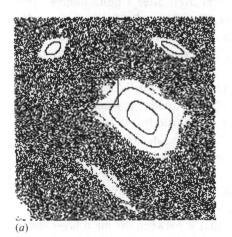
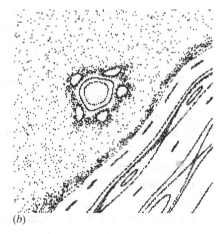

(a) (b)

consider a limited portion of phase space, a stochastic layer appears as an element of a sandwich-like pattern. However, its internal structure is of the 'stochastic sea' type. Therefore, the description of the structural properties of phase space always involves either a certain degree of localization of a given region and subsequent loss of information on the system as a whole, or concentration on certain large-scale properties of the region and a partial loss of information on the small scale.

In this chapter, we shall dwell on some extremely subtle properties of dynamic systems due to the transition from the state of stochastic layer to that of stochastic sea.

4.1 The border of global chaos

The standard mapping (3.4.8)

$$I_{n+1} = I_n - K \sin \theta_n$$
$$\theta_{n+1} = \theta_n + I_{n+1} \tag{4.1.1}$$

is the very model that enabled Green to determine the exact border of global chaos, i.e., the value of parameter $K = K_c$ that marks the formation of a stochastic sea [8] (Note 4.4).

Consider the mapping (4.1.1) on a torus, i.e.,

$$0 < I < 2\pi; \qquad 0 < \theta < 2\pi. \tag{4.1.2}$$

For $K = 0$ we have

$$I_n = \text{const} = I_0; \qquad \theta_n = \theta_0 + nI_0. \tag{4.1.3}$$

Equations (4.1.3) describe the family of invariant curves. When $I_0/2\pi$ is rational, an invariant curve closes onto itself after a finite number of steps of the mapping and is periodic. It consists of a finite number of points. On the contrary, when $I_0/2\pi$ is irrational, an invariant curve is everywhere densely filled by points of the trajectory, the latter ergodically filling the surface of a torus.

When $K \ll 1$, invariant curves exist by virtue of the KAM theory. Since the dimensionality of invariant tori is 2, they divide the phase space and, again according to the KAM theory, stochastic layers are sandwiched between invariant curves (see Sect. 2.3). Consider the main stochastic layer (Fig. 4.0.1). Invariant curves lie both inside and outside this. An increase in parameter K leads to the expansion of the stochastic layer. Beginning with the critical value K_c, all stochastic layers outside the main layer merge, giving birth to a stochastic sea. During this process, the destruction of invariant curves situated between the main layer and the one nearest to it takes place. Therefore, to define the value of K corresponding to stochastic sea formation, the value of K_c marking the destruction of the last remaining invariant curve must be determined. So far, the following estimate of this value has been used:

$$K_c \sim 1. \tag{4.1.4}$$

Now we shall discuss the possibility of determining the exact value of K_c.

Let us introduce the notion of the rotation number $\hat{\varphi}$ [14]. Consider a trajectory lying on a two-dimensional torus. In this case, $\hat{\varphi}$ is defined as the average angle of the inclination of trajectory divided by 2π. For example, in the case of (4.1.3) we simply have

$$\hat{\varphi} = I_0/2\pi. \tag{4.1.5}$$

Let $\varphi_1(t)$ and $\varphi_2(t)$ be two coordinates belonging to the trajectory on the torus. Then

$$\hat{\varphi} = \lim_{t \to \infty} \frac{\varphi_1(t)}{\varphi_2(t)}, \tag{4.1.6}$$

where symmetry in the choice of φ_1 and φ_2 is determined by the axis, in reference to which the computation of rotation number is performed. In the case of the mapping (4.1.1), let us define $\hat{\varphi}$ as follows:

$$\hat{\varphi} = \lim_{m \to \infty} \frac{1}{2\pi m} \sum_{n=1}^{m} (\theta_n - \theta_{n-1})$$

$$= \lim_{m \to \infty} \frac{1}{2\pi m} (\theta_m - \theta_0) = \lim_{m \to \infty} \frac{1}{2\pi m} \theta_m \tag{4.1.7}$$

in accordance with (4.1.6), if $\varphi_1(t)$ coincides with discrete time. If $\hat{\varphi}$ is rational,

$$\hat{\varphi} = \frac{Q_1}{Q_2}, \tag{4.1.8}$$

where Q_1, Q_2 are integers. Expression (4.1.8) signifies the fact that the trajectory closes on itself after Q_2 steps of the mapping, having completed Q_1 revolutions around the axis of the torus.

In the general case of KAM-tori, the rotation number $\hat{\varphi}$ is irrational. On the contrary, the tori with rational $\hat{\varphi}$ for $K = 0$ are resonant and must be destroyed if $K \neq 0$. Therefore, the problem of determining K_c can be formulated as the problem of determining the critical value of the rotation number $\hat{\varphi}_c$ for the last remaining invariant curve near the outer border of a stochastic layer.

Let us consider the following representation of the rotation number in the form of a continuous fraction:

$$\hat{\varphi} = \cfrac{1}{a_1 + \cfrac{1}{a_2 + \cdots}}$$

or, in a more compact notation:

$$\hat{\varphi} = [a_1, a_2, \ldots]. \tag{4.1.9}$$

For rational $\hat{\varphi}$, the sequence of numbers $a_1, a_2, \ldots$ is finite. For irrational $\hat{\varphi}$, this sequence is infinite. Cutting off the sequence at a certain a_Q, the following approximation

$$\hat{\varphi} \approx \hat{\varphi}_Q = [a_1, a_2, \ldots, a_Q] = \frac{Q_1}{Q} \tag{4.1.10}$$

is the most appropriate for any $Q' < Q$.

Now let us once more direct our attention to the mapping (4.1.1). For $K = 0$, let us consider the trajectories with a rational rotation angle (4.1.8). Since the step of the mapping is assumed to be unity, the perturbation incorporates the periods Q (Q is an integer), i.e., the frequencies $2\pi/Q$. In this case, it is possible to claim that for any Q the trajectories with rational $\hat{\varphi}$ (4.1.8) are in resonance with the perturbation. In their vicinity, separatrix cells are formed which correspond to a nonlinear resonance of the order of Q. Besides, the separatrices of this resonance are somewhat destroyed and dressed with a narrow stochastic layer (Fig. 4.1.1). The invariant curves nearest to the layer apparently have the rotation number $\hat{\varphi} \approx \hat{\varphi}_Q$, i.e., they are well approximated by equation (4.1.10). The reverse is also valid. If, for an invariant curve with the rotation number $\hat{\varphi}$, the approximation (4.1.10) turns out to be well fitting, this curve lies in the vicinity of a resonance of the Qth order.

What should be the value of $\hat{\varphi}$ for any finite approximation to be equally bad? According to Green's assumption, this value is the golden mean:

$$\varphi_c = [1, 1, \ldots] = \tfrac{1}{2}(\sqrt{5} - 1). \tag{4.1.11}$$

Fig. 4.1.1   Separatrix cells corresponding to a third-order resonance ($Q = 3$).

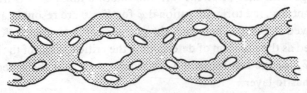

It is taken to be equal to the critical value. A numerical evaluation of K_c for (4.1.11) yields:

$$K_c = 0.9716 \cdots . \tag{4.1.12}$$

Despite a not-too-rigorous basis for the above reasoning, the numbers (4.1.11) and (4.1.12) turned out to be very reliable. Thus, our understanding of certain features of the onset of chaos was proved to be true.

In conclusion to this paragraph, let us stress that the KAM theory and the sandwich-like structure of phase space have been essential elements of all the steps in Green's method. It is this circumstance, as we shall see later, that restricts the method's universality (Note 4.5).

4.2 Percival's variational principle

There is one more important modification to the idea of studying invariant tori and invariant curves on the plane of mapping in a critical situation, when a chaotic zone lies nearby. This modification belongs to Percival [15–17]. It is the formulation of the variational principle of finding of invariant tori, should they exist.

Let a Hamiltonian system with N degrees of freedom be defined by a set of canonical variables (q, p), where q and p are N-dimensional vectors. Let us assume that there exists an N-dimensional invariant torus Σ. It can be defined by means of N frequencies:

$$\omega = (\omega_1, \ldots, \omega_N) \tag{4.2.1}$$

and, correspondingly, N angular variables:

$$\varphi = (\varphi_1, \ldots, \varphi_N). \tag{4.2.2}$$

If the trajectory lies on the surface of the invariant torus Σ, we have the following decomposition for the coordinates q on Σ:

$$q_\Sigma = \sum_m q_m \, e^{i(m,\varphi)} \tag{4.2.3}$$

where $m = (m_1, \ldots, m_N)$ is an N-dimensional vector consisting of positive and negative integers m_j and

$$(m, \varphi) = m_1\varphi_1 + \cdots + m_N\varphi_N. \tag{4.2.4}$$

The decomposition of the velocity vector $\dot{q}$ can be defined in a similar way:

$$\dot{q}_\Sigma = \sum_m \dot{q}_m \, e^{i(m,\varphi)}. \tag{4.2.5}$$

With the help of (4.2.1)–(4.2.3), the expression (4.2.5) can be rewritten in the following way:

$$\dot{q}_\Sigma = \left(\omega, \frac{\partial}{\partial\varphi}\right) q_\Sigma = \sum_m i(m, \omega) q_m e^{i(m,\varphi)}, \qquad (4.2.6)$$

where the following operator was introduced:

$$\left(\omega, \frac{\partial}{\partial\varphi}\right) = \omega_1 \frac{\partial}{\partial\varphi_1} + \cdots + \omega_N \frac{\partial}{\partial\varphi_N}. \qquad (4.2.7)$$

The expressions

$$q = q_\Sigma(\varphi), \qquad \dot{q} = \dot{q}_\Sigma(\varphi) = \left(\omega, \frac{\partial}{\partial\varphi} q_\Sigma\right) \qquad (4.2.8)$$

define the torus Σ in a parametric form.

Let $L(q, \dot{q})$ be the Lagrangian of a system. It has the usual relation to the Hamiltonian $H(p, q)$:

$$L(q, \dot{q}) = (p, \dot{q}) - H(p, q) \qquad (4.2.9)$$

provided the equation

$$\dot{q} = \frac{\partial H(p, q)}{\partial p} \qquad (4.2.10)$$

can be resolved relative to p. All the equations (4.2.5)–(4.2.10) are well known in mechanics [18]. Here their purpose is to show the transition from the Hamiltonian description of a system to its description by means of the functions q and $\dot{q}$ in (4.2.8) on the N-dimensional torus Σ.

Let us define a functional on the torus:

$$\mathscr{L}(\omega, \Sigma) = \frac{1}{(2\pi)^N} \int_0^{2\pi} d\varphi_1 \cdots d\varphi_N L(q_\Sigma(\varphi), \dot{q}_\Sigma(\varphi)), \qquad (4.2.11)$$

where $\dot{q}_\Sigma(\varphi)$ was defined in (4.2.5). The variation of torus Σ is a torus not very different from Σ which is defined by the following expressions:

$$q = q_\Sigma(\varphi) + \delta q(\varphi),$$

$$\dot{q} = \left(\omega, \frac{\partial}{\partial\varphi} q_\Sigma(\varphi)\right) + \left(\omega, \frac{\partial}{\partial\varphi} \delta q(\varphi)\right). \qquad (4.2.12)$$

The variation of functional (4.2.11) leads to an expression

$$\delta\mathscr{L}(\omega, \Sigma) = \frac{1}{(2\pi)^N} \int_0^{2\pi} d\varphi_1 \cdots d\varphi_N \left[\frac{\partial L}{\partial q} - \left(\omega, \frac{\partial}{\partial\varphi}\right) \frac{\partial L}{\partial \dot{q}}\right] \delta q. \qquad (4.2.13)$$

The stationarity condition of functional $\mathcal{L}(\omega, \Sigma)$ and the periodicity of q and $\dot{q}$ over φ leads to the Lagrangian equations

$$\frac{\partial L}{\partial q_i} = \left(\omega, \frac{\partial}{\partial \varphi}\right) \frac{\partial L}{\partial \dot{q}_i}, \qquad (i = 1, \ldots, N). \tag{4.2.14}$$

From (4.2.14) it immediately follows that if $q_\Sigma(\varphi)$ is the solution of the system (4.2.14), then

$$q_\Sigma(\varphi) = q_\Sigma(\omega t + \varphi_0),$$

$$\dot{q}_\Sigma(\varphi) = \left(\omega, \frac{\partial}{\partial \varphi}\right) q_\Sigma(\omega t + \varphi_0). \tag{4.2.15}$$

Percival's result can be expressed in the form of the following theorem.

The smooth torus Σ is the invariant torus of the system executing conditionally-periodic motion with the frequencies ω on its surface, if and only if it is a fixed point of the functional $\mathcal{L}(\omega, \Sigma)$.

From Percival's result there specifically follows that KAM-tori are solutions of the following equation:

$$\delta \mathcal{L}(\omega, \Sigma) = 0 \tag{4.2.16}$$

if the frequencies ω are highly incommensurate and the dynamical problem is considered in the region of applicability of the KAM theory (for the standard mapping (4.1.1), the last condition means that $K \ll 1$).

Let us again turn our attention to the standard mapping (4.1.1). The Hamiltonian (3.4.1) that enabled us to derive equations (3.4.8) and (4.1.1) has been already presented.

We can easily write down the Lagrangian and, making use of the expression (3.4.1) for

$$L = \tfrac{1}{2}\dot{\theta}^2 + K \cos \theta \sum_{n=-\infty}^{\infty} \delta(t/T - n) \tag{4.2.17}$$

the Lagrangian equation (4.2.14) acquires the following form:

$$\ddot{\theta} = -K \sin \theta \sum_{n=-\infty}^{\infty} \delta(t/T - n). \tag{4.2.18}$$

This agrees with the equations of motion (3.4.7) provided the substitution $\dot{\theta} = I$ has been taken into account, and after integration over the period of perturbation T results, as in Sect. 3.4, in the mapping (4.1.1). Let us write it in the following form:

$$\theta_{n+1} - 2\theta_n + \theta_{n-1} = -K \sin \theta_n = F(\theta_n). \tag{4.2.19}$$

Although at first sight the expressions (4.2.17) and (4.2.18) seem to lead to the same result as the Hamiltonian formalism, they now, however, have a new meaning that lies in the fact that equation (4.2.19) must have a periodic solution with frequency ω incommensurate with frequency $\nu = 2\pi/T$.

Variable θ, generally speaking, was not meant to be a periodical function. However, it can be presented in the following way:

$$\theta = \theta(\varphi) = \varphi + q(\varphi), \qquad (4.2.20)$$

where

$$\varphi = \omega t + \varphi_0 \qquad (4.2.21)$$

and $q(\varphi)$ is a periodic function of φ. By substituting (4.2.20) into equation (4.2.19), we get

$$q_{n+1} - 2q_n + q_{n-1} = -K \sin(\varphi + q_n(\varphi)) \equiv F_n(\varphi). \qquad (4.2.22)$$

Taking into account the fact that $q(\varphi)$ is a periodic function of φ, we can write the following expansions:

$$q_n(\varphi) = \sum_m q_n^{(m)} e^{im\varphi},$$

$$F_n(\varphi) = \sum_m F_n^{(m)} e^{im\varphi}. \qquad (4.2.23)$$

Now we perform the Fourier transform of the left- and right-hand sides of equation (4.2.22) and make use of the expressions (4.2.23). This finally gives us

$$q_n^{(m)} = \frac{F_n^{(m)}}{4 \sin^2(m\omega T)}, \qquad (4.2.24)$$

where the relation

$$q_{n+1}^{(m)} = q_n^{(m)} e^{im\omega T},$$

following from the definitions of θ_n and the formulas (4.2.20) and (4.2.21), is taken into account. Recall that the value of $\omega T/2\pi = \omega/\nu$ is irrational.

For numerical evaluation of the harmonics $q_n^{(m)}$, one has to select a finite number of them and perform an approximate evaluation of the amplitudes $F_n^{(m)}$, finally solving equations (4.2.24) with their help. An increase in parameter K, at a certain value K_c, results in divergence of the above method of finding the periodical orbits. By considering 15 harmonics in expansions (4.2.23), Percival got $K_c = 0.97161$ which coincides with (4.1.12). By these means, Percival's method made it possible to determine the critical parameter of perturbation at a fairly low cost. Although this procedure is not altogether unequivocal, since the divergence of the process of solving the system (4.2.22) may be connected with the appearance of small denominators, Percival's variational principle nevertheless makes it possible to arrive most naturally at a new notion, i.e., at the Cantorus (the term also belonging to Percival).

4.3 Cantori

We have considered above the problem of finding the conditionally-periodic solution of a given dynamic system. Initially, we are given a set of frequencies $\omega = (\omega_1, \ldots, \omega_N)$, where N is the number of degrees of freedom. This set is an arbitrary one and we are to find the invariant torus, around which the trajectory with frequencies ω is wound. As it turned out, the solution of this problem boils down to solving the Lagrangian equations (4.2.14). In the stability region their solutions are KAM-tori. There arises a new question: what will be the solution of equations (4.2.14) in that portion of phase space where the motion is stochastic? It has been answered only for several of the simplest cases [17, 19, 20].

Consider the equations of the standard mapping (4.2.19) at $K > 1$. In this case, there is a stochastic sea embracing numerous islands of stochasticity. There are no invariant tori in the stochastic sea (otherwise, diffusion in phase space would be impossible). However, the existence of a porous invariant torus is not ruled out. Percival named the solution of this type a Cantorus, since its cross-section is an invariant curve of a Cantor type. In other words, a Cantorus for (4.2.19) has the form of a cylinder, its cross-section forming a Cantor set. The solution of (4.2.19) can be presented in the following form:

$$\theta_n = 2\pi(n\nu + \nu_0) + f(n\nu + \nu_0) \tag{4.3.1}$$

where ν_0 is an arbitrary constant ($0 < \nu_0 < 1$) and f is a periodic function with the period $1/\nu$. In the case of a Cantorus, f has a complex discontinuous structure.

The following example of an exact computation of a Cantorus was given by Percival in [22]. Consider a mapping:

$$q_{n+1} - 2q_n + q_{n-1} = \frac{1}{a^2} s(q_n), \tag{4.3.2}$$

where $0 < q < 1$; the discontinuous function $s(q)$ is defined by the following formula:

$$s(q) = q - \tfrac{1}{2} - [q]$$

and $[q]$ denotes the integer part of q.

What makes (4.3.2) different from (4.2.19) is the fact that the periodic function $\sin \theta_n$ has been substituted by another periodic function $s(q)$. The exact solution of (4.3.2) in the case of sufficiently small a for Cantori has the following form:

$$q(\xi) = \xi - \frac{1}{(1 + 4a^2)^{1/2}} \sum_{m=-\infty}^{\infty} f^{-|m|} s(\xi + m\nu), \qquad (4.3.3)$$

where

$$\rho = 1 + \frac{1}{4a^2} [1 + (1 + 4a^2)^{1/2}]$$

and the values of q_n can be obtained from (4.3.3) provided we set $\xi = n\omega + \nu_0$.

The process of formation of the Cantori is as follows. At $K > K_c = 0.9716\cdots$, at least one rupture appears on an invariant curve (if the latter exists) in the case of mapping (4.2.19). Due to iterations of this rupture, an infinite number of new ruptures appear, which do not intersect and are steadily reducing in size. Thus, a Cantor set is formed which comprises an invariant curve being a cross-section of an invariant torus. The Cantori are unstable since small perturbations in the direction perpendicular to them are quickly increasing.

Being the invariant solutions of equation (4.2.19) with irrational rotation numbers, the Cantori abound in phase space and form barriers against the diffusion of particles in a stochastic sea. The smaller the gaps in a Cantorus, the more it slows down the diffusion. The barriers most impenetrable for diffusion lie close to the border of a stochasticity region. Having reached the neighbourhood of a Cantorus, a point designating the state of a system on the phase plane begins slowly moving along the Cantorus until it reaches a gap (Fig. 4.3.1) and gets an opportunity to pass right through the Cantorus (Note 4.6).

In the numerical analysis of a system's diffusion within a stochastic sea region, the Cantori manifest themselves as follows. At first, the random walk of a particle takes place in a narrow finite portion of phase space (Fig. 4.3.2a). This picture is characteristic, for example, for a standard mapping at $K = 1.1$ in Fig. 4.3.2a up to $t = 3646$ (the number of steps of the mapping). Motion is caught in a certain layer (see Fig. 4.3.2*a*). Further, the trajectory breaks through an invisible Cantorus quickly filling a far greater portion of phase space (Fig. 4.3.2*b*). Now its motion is limited by another Cantorus with large barriers and small gaps. This continues up to the time instant $t = 15\,268$ after which the next 'breakthrough' occurs and so on. In reality, the number of such spasmodic advancements is much greater, Fig. 4.3.2 depicting only the most conspicuous among them. With the increase of parameter K, the motion becomes more and more even, resembling the usual stochastic process. In the case of large K, the Cantori are very 'porous' and the effective length of the barriers becomes extremely small.

The complexity of the above process is due to the fact that it is an intermediate process and takes place in a border region already displaying the main characteristics of stochastic motion, but still showing traces of regular dynamics.

4.4 Hamiltonian intermittency

The border of chaos discussed in Sect. 4.1, apparently, is one of the simplest cases. It owes its existence to a specially chosen type of the model. Here we shall present a completely different picture of the border of chaos. The problem we are going to discuss first appeared as the result of attempts to realize the idea of stochastic Fermi acceleration in regular fields (Note 4.7).

First, let us turn our attention to certain qualitative considerations showing how the border of stochastic dynamics appears in the problems concerning the acceleration of particles. Let a particle travel in the field

Fig. 4.3.1   The passing of the trajectory through gaps in the Cantori.

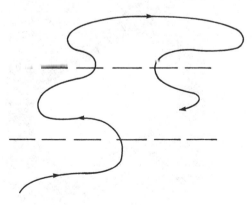

Fig. 4.3.2 Five consecutive pictures on the phase plane for a standard mapping at $K = 1.1$, corresponding to time instances $t = 3646$ (a); 15 268 (b); 24 245 (c); 29 122 (d); 34 984 (e). Each subsequent step is related to the 'breakthrough' of the trajectory through a Cantorus. Therefore, on the boundaries of the displayed stochastic sea regions there are Cantori which are invisible in the pictures.

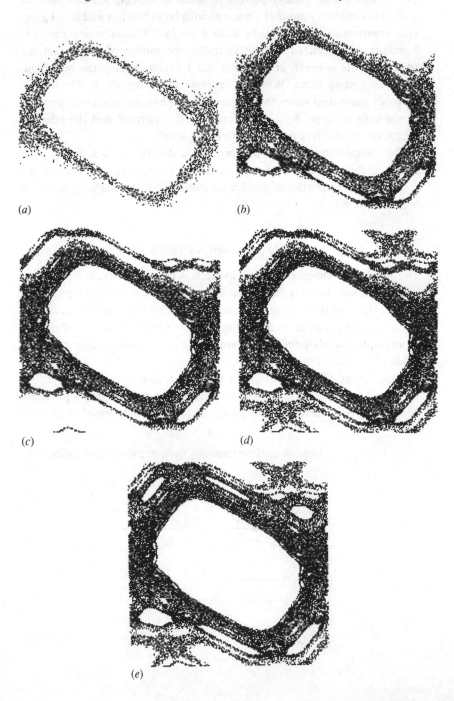

(a)

(b)

(c)

(d)

(e)

of a wave packet

$$\ddot{x} = \frac{e}{m} \sum_{n=-\infty}^{\infty} E_n \cos(k_n x - \omega_n t) \qquad (4.4.1)$$

in which

$$\Delta \omega = 2\pi / T \qquad (4.4.2)$$

is the characteristic distance between the frequencies ω_n, and

$$\Delta k = 2\pi / L \qquad (4.4.3)$$

is the characteristic distance between the wave numbers k_n of the wave packet.

Although the right-hand side of (4.4.1) is regular (non-random), nevertheless, under certain conditions discussed below, a particle's motion becomes stochastic and it starts accelerating. As the velocity of the particle grows, the time over which it covers the distance L becomes shorter. This means that the degree of the adiabaticity of motion increases and we should expect the appearance of the border of chaos. We may also expect that chaos, in this case, will be very weak as compared to the regular component of motion. Besides, the chaotic component of motion must have larger spatial and time scales in comparison with the high-frequency regular motion. In the theory of turbulence, this characteristic of motion is called intermittency [31, 32].

Usually, under the term 'intermittency' one understands space–time chaotic dynamics with a fairly well-pronounced space–time structure. We can most easily picture the intermittency in dissipative systems. The dynamics of a system in the vicinity of a stable limiting cycle is mostly determined by the spectrum of the cycle. If, for example, the cycle has only one period, the Fourier spectrum of a system in the vicinity of the cycle resembles the δ-function, its maximum corresponding to the cycle's frequency. If we have very weak chaos in the vicinity of a cycle, the system's spectrum also resembles the δ-like one. The evolution in time can be pictured as long regions of regular oscillations irregularly sewn together. Large regions of regular motion already imply a considerable degree of dynamic regularity. The spatial intermittency in the case of motion of the continuous medium looks much the same (Note 4.8).

A similar picture of motion very close to regular, but with very weak chaotic dynamics, is possible also in Hamiltonian systems. In this section we shall study the phenomenon of Hamiltonian intermittency in the problem of a particle's motion in the field of a wave packet (4.4.1). However, as we shall see later, the character of the phenomenon being described is fairly universal and is inherent in many problems which at first sight seem to have nothing in common [36].

We shall assume that the wave packet in (4.4.1) is very wide, i.e., that a large number of its harmonics have amplitudes E_n sufficiently different from zero. This property can be formally expressed if we set

$$E_n \approx E_0 \qquad (4.4.4)$$

for all n. Let us also assume

$$\omega_n \approx \omega_0 + n\,\Delta\omega; \qquad k_n \approx k_0 + n\,\Delta k. \qquad (4.4.5)$$

Introducing the group velocity of a packet

$$v_g = \Delta\omega / \Delta k, \qquad (4.4.6)$$

we can single out two limiting cases: $v > v_g$ and $v < v_g$. We shall consider the first of these, since the phenomenon of intermittency appears in the case of large particle velocities. At the condition $v \gg v_g$, from (4.4.5) it follows that we may assume $\omega_n \approx \omega_0$ for all n. By making use of this fact and equations (4.4.4) and (4.4.5), we get the following equation instead of (4.4.1)

$$\ddot{x} = \frac{e}{m} E_0 \sum_{n=-\infty}^{\infty} \cos(k_0 x - \omega_0 t + n\,\Delta kx)$$

$$= \frac{e}{m} E_0 \cos\theta \sum_{n=-\infty}^{\infty} \cos n\,\Delta kx, \qquad (4.4.7)$$

where we introduce a phase

$$\theta = k_0 x - \omega_0 t. \qquad (4.4.8)$$

But making use of the following relation

$$\sum_{n=-\infty}^{\infty} \cos n\,\Delta kx = L \sum_{n=-\infty}^{\infty} \delta(x - nL) \qquad (4.4.9)$$

and of notation (4.4.3), we can rewrite equation (4.4.7) as follows

$$\dot{v} = \frac{e}{m} E_0 L \cos\theta \sum_{n=-\infty}^{\infty} \delta(x - nL); \qquad v = \dot{x}. \qquad (4.4.10)$$

This notation reveals the physical meaning of the original problem, under the above assumptions (4.4.4), (4.4.5) and $v \gg v_g$. Provided the equations $x = x_n$ where

$$x_n = Ln = 2\pi n / \Delta k \qquad (4.4.11)$$

do not hold true, $\dot{v} = 0$, i.e. $v = $ const and the motion is free. At the moment when the coordinate of a particle reaches a value divisible by n, the particle is affected by an external force – it feels a kick. The kicks occur at the following moments: $\dots, t_n, t_{n-1}, \dots$ which, according to (4.4.11), can be found from the condition

$$\dots; x(t_n) = Ln; \quad x(t_{n+1}) = L(n \pm 1); \dots, \quad (4.4.12)$$

where the sign $\pm$ depends upon the direction of the velocity of a particle after the nth kick. Specifically, (4.4.12) gives us the time interval between two sequential collisions

$$\Delta t_n = |x(t_{n+1}) - x(t_n)|/|v_{n+1}| = L/|v_{n+1}| \quad (4.4.13)$$

where v_{n+1} is the particle's velocity after the nth kick. Let us construct a mapping relating the coordinate and velocity of a particle between two sequential mappings.

We introduce a new variable

$$w = \tfrac{1}{2}mv|v|, \quad (4.4.14)$$

the absolute value of which is equal to the energy of a particle. From (4.4.14) follows an expression for velocity v:

$$v = (2|w|/m)^{1/2} \operatorname{sgn} w. \quad (4.4.15)$$

By using definition (4.4.11) and the properties of δ-functions, we find from (4.4.10) and (4.4.14)

$$\dot{w} = eE_0L \cos \theta \sum_{n=-\infty}^{\infty} \delta(t - t_n). \quad (4.4.16)$$

Differentiating the expression for θ (4.4.8) and taking into account equation (4.4.15) we get

$$\dot{\theta} = k_0 v - \omega_0 = k_0(2|w|/m)^{1/2} \operatorname{sgn} w - \omega_0. \quad (4.4.17)$$

The set of equations (4.4.16) and (4.4.17) with respect to the new variables (w, θ) is closed. These variables also form a canonically conjugate pair, since the following relation holds true for them

$$\frac{\partial \dot{w}}{\partial w} + \frac{\partial \dot{\theta}}{\partial \theta} = 0.$$

In order to construct a mapping on the plane (w, θ), we must notice that a sequence of time instants $t_0, t_1, \dots, t_n, \dots$ in each of which a particle feels a kick, is monotonically increasing, i.e., the inequality $t_n > t_k$ holds true if $n > k$. However, the sequence of coordinates $x_0, x_1, \dots, x_n, \dots$

marked by the kicks is not monotonic, since the direction of a particle's velocity may vary with the kick. This is reflected in (4.4.13).

Let us denote

$$w_n = w(t_n - 0); \qquad \theta_n = \theta(t_n - 0). \tag{4.4.18}$$

In accordance with the equations of motion (4.4.16) and (4.4.17), in the course of a kick, variable θ remains continuous, i.e.

$$\theta(t_n - 0) = \theta(t_n + 0), \tag{4.4.19}$$

while variable w exhibits a discontinuity. By taking into account the property (4.4.19), we obtain from (4.4.16) and (4.4.17)

$$w(t_n + 0) - w(t_n - 0) = eE_0 L \cos \theta_n,$$
$$\theta(t_n + 0) - \theta(t_n - 0) = 0. \tag{4.4.20}$$

During the time interval $(t_n + 0, t_{n+1} - 0)$, the value of w remains unchanged and, consequently,

$$
\begin{aligned}
w(t_{n+1} - 0) = w_{n+1} &= w(t_n + 0) \\
&= w(t_n - 0) + eE_0 L \cos \theta_n \\
&= w_n + eE_0 L \cos \theta_n.
\end{aligned} \tag{4.4.21}
$$

For the same reason, $\dot{\theta}$ is a constant during the same time interval and (4.4.20) yields

$$\theta(t_{n+1} - 0) = \theta_{n+1} = \theta(t_n + 0) + \dot{\theta}(t_n + 0)\,\Delta t_n \tag{4.4.22}$$

where Δt_n is defined by equation (4.4.13). From (4.4.21) and (4.4.22) follow the final equations defining the $\hat{L}$-mapping

$$\hat{L}: \begin{cases} w_{n+1} = w_n + eE_0 L \cos \theta_n \\ \theta_{n+1} = \theta_n + k_0 L \operatorname{sgn} w_{n+1} - \omega_0 L (m/2|w_{n+1}|)^{1/2} \pmod{2\pi} \end{cases} \tag{4.4.23}$$

It is convenient to rewrite equations (4.4.23) by introducing the new normalized values

$$u = 2w/m\omega_0^2 L^2; \qquad y = \theta/2\pi - 1/4. \tag{4.4.24}$$

This yields

$$\hat{L}: \begin{cases} u_{n+1} = u_n + Q \sin 2\pi y_n \\ y_{n+1} = y_n - \dfrac{1}{2\pi|u_{n+1}|^{1/2}} + \tilde{y} \operatorname{sgn} u_{n+1} \pmod{1} \end{cases} \tag{4.4.25}$$

where $\tilde{y} = k_0 L/2\pi$ is a constant shift in phase and

$$Q = 2eE_0/m\omega_0^2 L. \tag{4.4.26}$$

Q has the meaning of the ratio of the energy change of a particle in the course of a single collision ($mv\delta v \sim eE_0 L$) to the energy of the high-frequency oscillations of the particle ($\sim m\omega_0^2 L^2$).

At fairly small energy values $|u|$, the changes in phase with each kick are large and the trajectory of a particle becomes stochastic. The conditions for the onset of chaos can be roughly expressed in the following form

$$K = \left| \frac{\delta y_{n+1}}{\delta y_n} - 1 \right| = \frac{Q}{2|u|^{3/2}} |\cos 2\pi y| \geqslant 1. \qquad (4.4.27)$$

The phase portrait of a system at $Q \ll 1$ is presented in Fig. 4.4.1. Stability islands, where the trajectories are regular, are determined by multiplier $\cos 2\pi y$ in (4.4.27), due to which the inequality does not hold true for a certain range of the values of y.

The main characteristic of the $\hat{L}$-mapping which distinguishes it from the standard mapping (Sect. 3.4) is the fact that, in the equation for y defining the change in phase, the action u is present in the denominator and not in the numerator. In the case of $Q \ll 1$, if originally the energy was low ($|u| \ll 1$), its changes $\delta u \sim Q$ are also small, so that on a long interval of time the value of $|u|$ may continue to be small. However, if $Q \gg 1$ the picture is different. Even one kick gives us $|u| \sim Q$, even if the initial value $|u| \ll 1$. This immediately yields

$$\delta y = |y_{n+1} - y_n| \sim \frac{1}{2\pi Q^{1/2}} \ll 1. \qquad (4.4.28)$$

Fig. 4.4.1 The phase plane of $\hat{L}$-mapping at $Q = 10^{-3}$, $\tilde{y} = 0$.

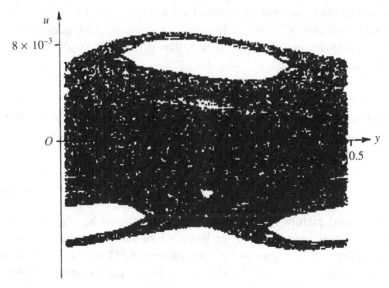

Condition (4.4.28) means an extremely slow adiabatic change in phase y from one collision to another. Consider the following range of values

$$|u| \gg Q \gg 1.$$

Within this range, a change over one step of the mapping is

$$|\delta u| \sim Q \ll |u|. \tag{4.4.29}$$

The conditions (4.4.28) and (4.4.29) enable us to approximate the $\hat{L}$-mapping (4.4.25) with a good accuracy by the following differential equations

$$\frac{du(n)}{dn} = Q \sin 2\pi y(n)$$

$$\frac{dy(n)}{dn} = -\frac{1}{2\pi |u(n)|^{1/2}} + \tilde{y}. \tag{4.4.30}$$

The set (4.4.30) has a regular non-random solution

$$u^{1/2} = \tfrac{1}{2} Q(1 + \cos 2\pi y) + C, \qquad (u > 0) \tag{4.4.31}$$

defining an invariant curve where C is an arbitrary constant and, for simplicity, it is assumed $\tilde{y} = 0$. However, equation (4.4.31) does not describe a real situation, since there are moments of time such that the original system (4.4.30) does not make sense. In the case of large values of Q, a change in the phase y over one step is small, according to (4.4.28). The value of u in the $\hat{L}$-mapping (4.4.25) changes equally slowly. The process of the slow changes of u and y is described by the curve (4.4.31) to which all points of the trajectory (u, y) belong. This goes on until $\sin 2\pi y$ approaches zero. The alternation of its sign results in a very rapid change

$$\delta u = |u_{n+1} - u_n| = Q|\sin 2\pi y_n|.$$

It is this subtle effect that causes a small stochastization of a particle's dynamics. The number of steps n_0, after which a non-adiabatic change in the phase y occurs, can be easily evaluated by means of equation (4.4.28)

$$n_0 = 2\pi Q^{1/2} \gg 1. \tag{4.4.32}$$

The general picture of this motion is clearly shown by Fig. 4.4.2. The phase plane is covered mostly by large groups with the number of points $\approx 2n_0$. Each of these groups lies on a curve (4.4.31) with a certain value of the constant C. As the difference $|u_{n+1} - u_n|$ becomes negative, equations (4.4.30) loose their applicability. However, later they become applicable once again, although their solution is a curve (4.4.31), this time with another constant C. The points where the trajectory jumps over to another curve in the family of (4.4.31), are clearly shown in Fig. 4.4.2 near the Oy axis at $u = 0$. Actually, this is a good presentation of the process of Hamiltonian intermittency on the phase plane.

A family of curves

$$u = u(y, C) \qquad (4.4.33)$$

covers the phase plane, while the trajectory of a system is a set of curves (4.4.33) sewn together at the different randomly varying values of the constant C. Now we are able to inquire about the limiting value of C_0 that defines the topmost invariant curve in Fig. 4.4.2, which is the border of the stochastic dynamics lying below. A numerical analysis [36] has shown that

$$C_0 \approx 0.5 Q^{1/3}.$$

In general, the border of stochasticity can be approximated by the following curve

$$u = \tfrac{1}{4} Q^2 [1 + \cos(2\pi y + \delta_0) + Q^{-2/3}]^2 \qquad (4.4.34)$$

Fig. 4.4.2 The phase plane of $\hat{L}$-mapping at $Q = 10$ and $\tilde{y} = 0$ corresponds to the case of Hamiltonian intermittency.

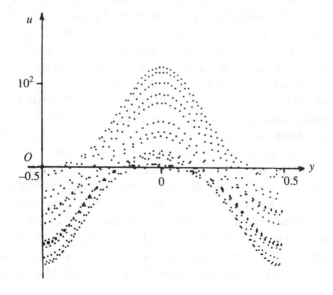

where the shift in phase $\delta_0 \to 0$ at $Q \to \infty$. Let us note that for $Q \ll 1$, equation (4.4.34) is a good approximation of the stochasticity border obtained from (4.4.26). Therefore, we may expect that it will have a more universal character than we assumed during its derivation.

The above phenomenon of stochastic dynamics in the case of intermittency also results in the acceleration of a particle, although this process is much slower than one taking place at small values of Q. The energy gained in the course of acceleration can be very high. We can estimate it from (4.4.34),

$$u_{max} \sim Q^2. \tag{4.4.35}$$

In actuality, the process of acceleration due to stochastic intermittency is highly sensitive towards the details of the mechanism of the breaking-up of adiabaticity. As will be shown later, estimation (4.4.35) is not limiting, while the above picture of intermittency is not unique.

Let us consider the case of $\tilde{y} \neq 0$. Let us also assume $0 < \tilde{y} < 1$. Then from (4.4.30) the existence of two more singular points with the coordinates

$$(y_1 = 0, \ u_1 = 1/4\pi^2\tilde{y}^2);$$

$$(y_2 = \tfrac{1}{2}, \ u_2 = 1/4\pi^2\tilde{y}^2)$$

follows. The first one is hyperbolic, while the second one is elliptic. If the point (y_1, u_1) does not fall into the stochasticity region, the phase portrait of the system is analogous to the one depicted in Fig. 4.4.2. However, if the hyperbolic point has a sufficiently low location (which depends upon the value of $\tilde{y}$), a narrow stochastic channel branches out and goes parallel to the separatrix whiskers passing through the point (y_1, u_1) (see Fig. 4.4.3). The formed region of chaotic dynamics enables far greater accelerations as compared to ones predicted by estimation (4.4.35). Thus, for the example in Fig. 4.4.3, which is already in the case of $\tilde{y}$ slightly different from zero ($\tilde{y} = 0.007$), $u_{max} \sim Q^3$. The high sensitivity of acceleration towards the variations of parameter $\tilde{y}$ is accompanied by an abrupt bifurcational change in the character of intermittency. Now it includes two families of regular trajectories: trajectories of the type already described and trajectories parallel to the separatrices.

The above example shows that the border of chaos defined in Sect. 4.1 for a standard mapping is not universal. It also shows the relation between the rotation number of the invariant curve and the golden mean. This becomes evident from equations (4.4.31) and (4.4.34). What is more, the destruction of the topmost invariant curve might come, not as a result of the transition towards global chaos, but as the result of a merging of only two stochastic regions (Fig. 4.4.4). The destruction of an invariant curve near the border of chaos is connected with the implantation of a saddle in the region of stochastic dynamics.

Hamiltonian intermittency covers a large class of physical problems which might include more than the problems in which the change in phase is proportional to the inverse of a certain power of action. The example we are going to discuss in the next section reveals certain additional cases of the appearance of weak chaos.

Fig. 4.4.3 The phase plane of $\hat{L}$-mapping at $Q = 10$ and $\tilde{y} = 0.07$ demonstrates a strong change in the intermittency picture.

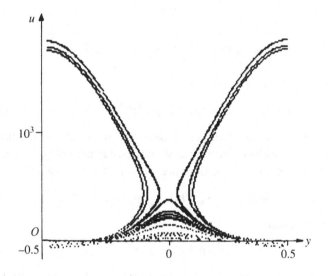

Fig. 4.4.4 The phase portrait for the case of $Q = 10$, $\tilde{y} = 0.003$ (a) and $\tilde{y} = 0.008$ (b).

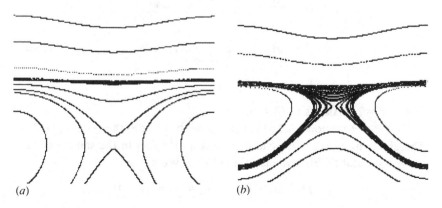

(a) (b)

4.5 The acceleration of relativistic particles

Let us once more turn our attention to equation (4.4.1). This describes the motion of particles in the field of a wave packet. If the acceleration of particles is sufficiently large, the non-relativistic approximation turns out to be incorrect and we should turn to the dynamics of relativistic particles (Note 4.9).

In the relativistic case, we write the equations of motion in the following form [41]

$$\dot{p} = -eE(x, t); \qquad \dot{x} = pc^2/\mathcal{E}, \qquad (4.5.1)$$

where

$$E(x, t) = \sum_{n=-\infty}^{\infty} E_n \sin(k_n x - \omega_n t) \qquad (4.5.2)$$

and $\mathcal{E}$ is the energy of a particle:

$$\mathcal{E} = (m^2 c^4 + p^2 c^2)^{1/2}. \qquad (4.5.3)$$

In contrast to the case discussed in Sect. 4.4, we shall assume that the wave packet is characterized by a single spatial period with the wave number k_0. In accordance with expressions (4.4.4) and (4.4.5), we set

$$E_n = E_0; \qquad k_n = k_0; \qquad \omega_n = \omega_0 + n\,\Delta\omega. \qquad (4.5.4)$$

Denoting as before

$$\theta = k_0 x - \omega_0 t \qquad (4.5.5)$$

we can easily obtain after substituting (4.5.4) and (4.5.5) into (4.5.1) and (4.5.2)

$$\dot{p} = -eE_0 \sin \theta \sum_{n=-\infty}^{\infty} \cos n\,\Delta\omega t$$

$$= -eE_0 T \sin \theta \sum_{n=-\infty}^{\infty} \delta(t - nT); \qquad (4.5.6)$$

$$\dot{\theta} = (k_0 pc^2/\mathcal{E}) - \omega_0$$

where $T = 2\pi/\Delta\omega$, as in (4.4.2).

The set of equations (4.5.6) has the same structure as the system (3.4.7). Therefore, in our case we can construct a mapping in the same way as for the standard Chirikov mapping (3.4.8). We assume

$$p_n = p(t = nT - 0); \qquad \theta_n = \theta(t = nT - 0).$$

The integration of (4.5.6) over a time interval $(nT-0, (n+1)T-0)$ yields

$$I_{n+1} = I_n - K \sin \theta_n,$$

$$\theta_{n+1} = \theta_n + \frac{I_{n+1}}{(1+I_{n+1}^2/\tau_k^2)^{1/2}} - \omega_0 T,$$

(4.5.7)

where the following dimensionless momentum is introduced

$$I = pk_0 T/m = \tau_k p/mc,$$

(4.5.8)

parameter K is of the same form as in Sect. 3.4

$$K = (e/m)E_0 k_0 T^2 \equiv \Omega_0^2 T^2,$$

(4.5.9)

and

$$\tau_k = k_0 cT.$$

(4.5.10)

Even a superficial comparison of the mapping (4.5.7) to the $\hat{L}$-mapping (4.4.25) detects a certain similarity between them. Let us present several limiting cases of equations (4.5.7).

As follows from (4.5.8), to a non-relativistic limit there corresponds a condition $I/\tau_k \to 0$. The mapping (4.5.7) turns into a standard mapping. In the case of an ultra-relativistic limit, on the contrary, $I/\tau_k \to \infty$ and from (4.5.6), by retaining the first terms in the expansion in τ_k/I, we get

$$I_{n+1} = I_n - K \sin \theta_n,$$

$$\theta_{n+1} = \theta_n + \tau_k \operatorname{sgn} I_{n+1} \left(1 - \frac{\tau_k^2}{2I_{n+1}^2}\right) - \omega_0 T \quad (\text{mod } 2\pi).$$

(4.5.11)

If $\tau_k = 2\pi m$ (where m is an integer), (4.5.11) transfers into the following equations

$$I_{n+1} = I_n - K \sin \theta_n,$$

$$\theta_{n+1} = \theta_n - \frac{\tau_k^3 \operatorname{sgn} I_{n+1}}{2I_{n+1}^2} - \omega_0 T \quad (\text{mod } 2\pi),$$

(4.5.12)

which are very close to (4.4.25). The difference lies only in the power of the action which stands in the denominator of the nonlinear frequency in the equation of phase y.

In the case of arbitrary $\tau_k \neq 2\pi m$ (integer m) in (4.5.11), the minor term τ_k^2/I^2 makes a minor contribution to the variations of phase θ. However, it plays an important role in determining the local instability, since it defines the derivative $\partial \theta_{n+1}/\partial I_n$.

Let us present several numerical illustrations of the relativistic mapping (4.5.7). Figure 4.5.1 shows its phase portrait at certain moderate values of parameters (not-too-strong relativity and not-too-weak instability). The stochasticity region has the usual set of resonance islands. Outside it lie the invariant curves.

Fig. 4.5.1 The phase portrait of the relativistic mapping (4.5.7). The intervals of changes are: horizontal $\theta \in (-\pi, \pi)$; vertical $I \in (3\pi, -3\pi)$. The values of parameters are: $K = 3$; $\tau_k = 4$; $\omega_0 T = 0$.

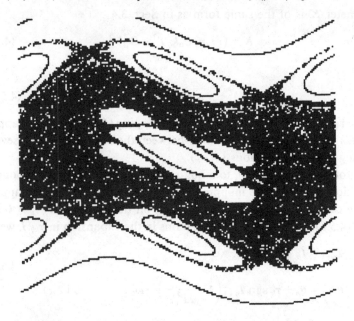

Fig. 4.5.2 The region of chaos for a relativistic mapping in the case of a strong intermittency. The intervals of changes are: horizontal $\theta \in (-\pi, \pi)$, vertical $I \in (-1500\pi, 1500\pi)$. The values of parameters are: $K = 200$, $\tau_k = 6$, $\omega_0 T = 0$. The computation times (the number of steps in the mapping) are: (a) 979; (b) 5×10^4.

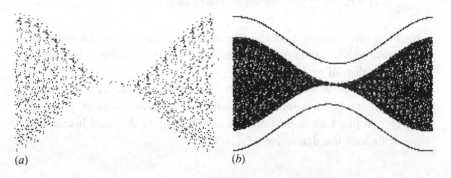

(a)   (b)

As the parameter K increases, a strong intermittency occurs. The points fall onto a regular curve which has irregular displacements only in the vicinity of values of abscissae (Fig. 4.5.2). To obtain a more graphic picture of intermittency, a moderate computational time (about 10^3 iterations) was chosen for Fig. 4.5.2a. In the case of intermittency, the regular curves comprising a trajectory can be found from equations

$$\frac{dI}{dn} = -K \sin \theta,$$

$$\frac{d\theta}{dn} = \frac{I}{(1+I^2/\tau_k^2)^{1/2}} - \omega_0 T. \qquad (4.5.13)$$

In the case of $\omega_0 T = 0$ from (4.5.13) we have

$$I^2 = -\tau_k^2 + (K/\tau_k)^2(\cos \theta + \text{const})^2 \qquad (4.5.14)$$

where const is an integration constant. By its appropriate choice we can get the topmost invariant curve near the border of chaos.

Let us give two more examples of the phase plane of the relativistic mapping (4.5.7) near the resonance values of τ_k. In Fig. 4.5.3, values of

Fig. 4.5.3 The phase portrait of a relativistic mapping in the vicinity of the resonance value $\tau_k = 2\pi$. The intervals of changes are: horizontal $\theta \in (-\pi, \pi)$; vertical $I \in (-10\pi, 10\pi)$. The values of parameters are: $K = 6$, $\tau_k = 6.28$, $\omega_0 T = 0$.

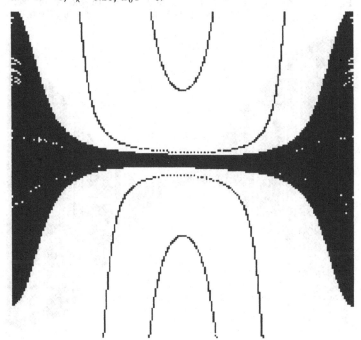

Fig. 4.5.4   As Fig. 4.5.3, but in conditions of strong intermittency: $\theta \in (-\pi, \pi)$; $I \in (-5 \times 10^4 \pi, 5 \times 10^4 \pi)$, $K = 19$, $\tau_k = 12.566$, $\omega_0 T = 0$.

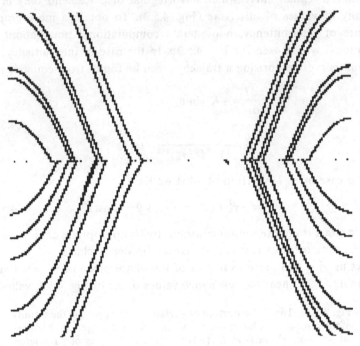

Fig. 4.5.5   As Fig. 4.5.2, but at a large value of parameter K: $\theta \in (-\pi, \pi)$, $I \in (-5 \times 10^4 \pi, 5 \times 10^4 \pi)$, $K = 10^4$; $\tau_k = 5.5$, $\omega_0 T = 0$.

K are moderate. However, energy levels reached by an accelerating particle are exceedingly high. The highest energies are concentrated near the phases $\theta = \pm \pi$. In Fig. 4.5.4 the value of K is large and the picture of intermittency is very clear. The action reaches values of 10^6. The mechanism of stochastic acceleration in the case of a strong intermittency has little in common with the mechanism of Fermi acceleration of a diffusion type. Specifically, for very large values of $K \gg \tau_k^2$, as shown by (4.5.14), the character of the intermittency picture may change (Fig. 4.5.5).

5
The stochastic web

The discovery of chaos in dynamic systems has given rise to a number of questions of a fundamental nature concerning the conditions of the onset of chaos and its properties. There are also questions of a different nature which aim not only at understanding the phenomenon of chaos on the whole but at a more general understanding of the global properties of dynamic systems. Therefore, the question as to whether chaos can be irremovable acquires a new dimension. A positive answer would mean that such systems in principle cannot be totally integrable. Therefore, the theory of the stochastic layer as discussed in chapter 3, in actuality contains much more information: in the general case, any dynamic system has an irremovable region of stochasticity in phase space. Let us repeat briefly why that is so.

Let us present the Hamiltonian of a system in the following form:

$$H = H_0(I) + \varepsilon V(I, \theta; t)$$

where I, θ are N_0-dimensional vectors. Such a system possesses $N = N_0 + \frac{1}{2}$ degrees of freedom, the value $\frac{1}{2}$ having to do with the variable t (time). In the general case, a part of the Hamiltonian H_0 has separatrices so that, in destroying them, the perturbation εV forms stochastic layers for any ε. Even when $N_0 = 1$, perturbations periodic in t yield the same result. Even though the existence of a stochasticity region thus becomes a universal feature of dynamic systems, this does not necessarily imply the existence of strong instability in the system. This statement is based on the fact that regions of chaos are very narrow. Therefore, the future form of a stochastic trajectory depends upon how the regions of chaos merge or, in other words, upon the topology of weak chaos in phase space.

The merging of all stochastic layers in phase space may form a single network – a stochastic web. The random walk of a particle along this web can take it arbitrarily far. Thus, the existence of a web means a qualitatively new manifestation of chaos: the universal mechanism of irremovable diffusion in phase space. In fact, this is only one of the consequences. The other, no less important, has to do with the geometry of the web. In the present chapter we shall see how a stochastic web appears.

5.1 KAM-tori and Arnold diffusion

In Sect. 2.3 we discussed the KAM theory which defines the conditions of the conservation of invariant tori in a Hamiltonian system affected by a small perturbation. According to one of the key concepts of this theory, the measure of destroyed tori is small and, what is more, small regions containing destroyed tori are clamped between invariant tori (Fig. 5.1.1). The same situation for a standard mapping with the number of degrees of freedom $N = \frac{3}{2}$ is shown in Fig. 3.4.1. The situation should be very different if there existed tori which could cross a certain network of channels with stochastic dynamics inside. However, for purely topological reasons such intersection is possible only for $N > 2$. The phenomenon of universal diffusion along the web thus formed at $N > 2$ was discovered by Arnold [1]. The mechanism of the formation of such a web is as follows.

First, we consider the case of $N = 2$ (the case of $N = \frac{3}{2}$ is examined similarly) and write the Hamiltonian of a system in the usual form:

$$H = H_0(I_1, \ldots, I_N) + \varepsilon V(I_1, \theta_1; \ldots; I_N, \theta_N). \qquad (5.1.1)$$

Fig. 5.1.1 The regions of destroyed tori are clamped between the invariant tori.

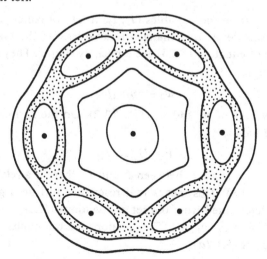

We introduce the frequencies of unperturbed motion on an N-dimensional torus:

$$\omega_j = \frac{\partial H_0}{\partial I_j} \qquad (j = 1, \ldots, N). \qquad (5.1.2)$$

Each frequency is, generally speaking, a function of all actions:

$$\omega_j = \omega_j(I_1, \ldots, I_N) \qquad (j = 1, \ldots, N). \qquad (5.1.3)$$

Let us single out the resonance tori defined by the following equation:

$$\sum_{j=1}^{N} n_j \omega_j = 0 \qquad (5.1.4)$$

where n_j are integer numbers. For every set of integers $(n_1, \ldots, n_N)$ there is a set of solutions $I^{(0)} \equiv (I_1^{(0)}, \ldots, I_N^{(0)})$. Each solution $I^{(0)}$ defines a resonance torus. It involves, on the one hand, the formation of a separatrix loop with a width $\sim \varepsilon^{1/2}$ and, on the other hand, its destruction and the formation of a stochastic layer in its place under the effect of various non-resonance terms in (5.1.1). This process was described at length in Sects. 2.1, 2.2, and 3.5. Now let us look at the same phenomenon from a somewhat different angle.

For $N = 2$, the condition (5.1.4) takes the following form:

$$n_1 \omega_1 + n_2 \omega_2 = 0. \qquad (5.1.5)$$

According to (5.1.3), each frequency depends upon two actions (I_1, I_2). Further, in accordance with the KAM theory we suggest the non-degeneracy of the system of frequencies (see equation (2.3.3)):

$$\det \left| \frac{\partial^2 H_0}{\partial I_j \, \partial I_k} \right| \neq 0. \qquad (5.1.6)$$

Let us transfer from the variables (I_1, I_2) to the variables (ω_1, ω_2). The formulas (5.1.2) define this change of variables. On the plane (ω_1, ω_2) the solutions of equation (5.1.5) are especially simple. They are a family of straight lines:

$$\omega_2 = -n_1 \omega_1 / n_2 \qquad (5.1.7)$$

for various values of n_1 and n_2 (Fig. 5.1.2). On the surface of a given energy E we have:

$$E = H_0(I_1, I_2). \qquad (5.1.8)$$

Therefore, on this surface, between I_1 and I_2, there is a relation defined by equation (5.1.8). This means that resonance tori on a given energy surface are determined from the set of equations (5.1.7) and (5.1.8) for each pair of integers (n_1, n_2). The solutions are the points on the plane (ω_1, ω_2) (see Fig. 5.1.2a).

Each of these points corresponds to a resonance and each of these resonances corresponds to a stochastic layer which creates a small region of chaos on the plane (ω_1, ω_2). When $\varepsilon \to 0$, the sizes of these regions tend to zero and they do not intersect.

When $N > 2$, everything is different. For example, if $N = 3$, instead of (5.1.5) we have:

$$n_1\omega_1 + n_2\omega_2 + n_3\omega_3 = 0, \qquad (5.1.9)$$

where each pair of frequencies depends on three actions (I_1, I_2, I_3). In the space $(\omega_1, \omega_2, \omega_3)$, equation (5.1.9) defines a family of surfaces. On the surface of energy E we have:

$$E = H_0(I_1, I_2, I_3) = \tilde{H}_0(\omega_1, \omega_2, \omega_3), \qquad (5.1.10)$$

where the function $\tilde{H}_0$ is obtained from H_0 by a change of variables. This is also the equation of a surface intersecting surfaces (5.1.9) along certain curves. Thus, resonance tori share some common regions along the curves which are the solutions of two equations, (5.1.9) and (5.1.10) with respect to three variables. A small perturbation $\sim\varepsilon$ dresses these curves with a stochastic layer of thickness $\sim\varepsilon^{1/2}$. This is what we call a stochastic web (Fig. 5.1.2b).

For small ε, the structure of the web is determined by only the two following equations:

$$E = H_0(I_1, \ldots, I_N),$$
$$n_1 \frac{\partial H_0}{\partial I_1} + \cdots + n_N \frac{\partial H_0}{\partial I_N} = 0 \qquad (5.1.11)$$

Fig. 5.1.2   The stochasticity regions at (a) $N = 2$ and (b) $N > 2$.

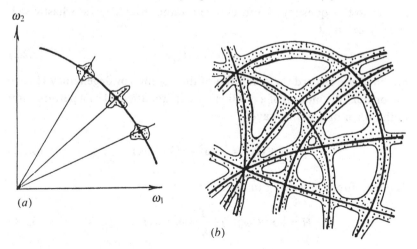

(a)

(b)

which are independent of ε. Therefore, a certain approximate form of the web has nothing to do with the perturbation in (5.1.1) and is determined by certain structural properties of dynamic systems using the Hamiltonian. A stochastic layer in this approximate pattern is irremovable since it exists at arbitrarily small ε. It is accompanied by irremovable diffusion in a system, the initial conditions of which happened to lie in the web's domain. This is Arnold's diffusion (Note 5.1).

5.2 Weak chaos and the stochastic web

We already know about the situation where an arbitrarily small perturbation results in the formation of such regions in phase space in which the dynamics of a system is stochastic. Examples of such regions are stochastic layers (for $N > 1$) and a stochastic web (for $N > 2$). The smaller the perturbation, the smaller the regions of chaos. However, chaos becomes irremovable if certain general conditions concerning the dimensionality of a system are satisfied. Loosely speaking, this case of very small regions of stochastic dynamics may be called a case of weak chaos. Regions of chaos, no matter how small, imply that a system is, in principle, nonintegrable.

Weak chaos manifests itself in the form of stochastic layers or a stochastic web. The KAM theory implies that a stochastic web exists at $N > 2$, i.e., that a system, in which it appears, has the following minimal dimensionality:

$$\min N = 2\tfrac{1}{2}. \tag{5.2.1}$$

Can (5.2.1) be reduced in value? A series of works [7–11] gives a positive answer to this question. To formulate it rather roughly, a stochastic web already exists at

$$\min N = 1\tfrac{1}{2} \tag{5.2.2}$$

provided we discard the condition of the absence of degeneracy (5.1.6).

Consider the motion of a nonlinear oscillator affected by a perturbation in the form of a plane wave:

$$\ddot{x} + \omega_0^2 x = \frac{1}{k}\,\varepsilon\omega_0^2 \sin(kx - vt). \tag{5.2.3}$$

The Hamiltonian leading to equation (5.2.3) is:

$$H = \tfrac{1}{2}(\dot{x}^2 + \omega_0^2 x^2) + \frac{1}{k^2}\,\varepsilon\omega_0^2 \cos(kx - vt). \tag{5.2.4}$$

The left-hand side of equation (5.2.3) describes very simple dynamics of linear oscillations. All the non-trivial types of motion arise from the perturbation and its interaction with the unperturbed motion. However, the current situation may be very different from the one defined by the KAM theory.

To make sure of this, let us transfer in (5.2.4) to new action-angle variables. For the variables $(x, \dot{x})$ let us set:

$$x = (2I_x/\omega_0)^{1/2} \sin \varphi_x,$$
$$\dot{x} = (2\omega_0 I_x)^{1/2} \cos \varphi_x. \tag{5.2.5}$$

Expressed in these variables,

$$H = \omega_0 I_x + \varepsilon V(I_x, \varphi_x; t),$$

$$V(I_x, \varphi_x; t) = \frac{1}{k^2} \omega_0^2 \cos\left[k \left(\frac{2I_x}{\omega_0}\right)^{1/2} \sin \varphi_x - \nu t \right]. \tag{5.2.6}$$

The unperturbed part of the Hamiltonian $H_0 = \omega_0 I_x$ does not satisfy the non-degeneracy condition (5.1.6). Therefore, in the case of a resonance

$$n\omega_0 = \nu \tag{5.2.7}$$

where n is an integer, the amplitude of an oscillator can greatly increase. There is nonlinearity and it is possible to escape from the resonance only due to the perturbation. Further we shall see that this situation is common to all systems close to linear.

The problem (5.2.3) has an extremely graphic interpretation. It is equivalent to the motion of a particle in a constant magnetic field B_0 and in the field of a plane wave travelling perpendicularly to the magnetic field. Let us write the equation of motion of a particle for this case in the following form:

$$\ddot{r} = \frac{e}{mc} [\dot{r}, B_0] + \frac{e}{m} E_0 \sin(kr - \nu t). \tag{5.2.8}$$

Let us assume that B_0 is directed along the z-axis, vector r lies in the plane (x, y), directing the vectors k and E_0 along the x-axis (a longitudinal wave). Then from (5.2.8) it follows that

$$\ddot{x} = \omega_0 \dot{y} + \frac{1}{k} \varepsilon \omega_0^2 \sin(kx - \nu t),$$
$$\ddot{y} = -\omega_0 \dot{x}, \tag{5.2.9}$$

where

$$\omega_0 = eB_0/mc, \qquad \varepsilon \omega_0^2 = eE_0 k/m.$$

From (5.2.9) the existence of an integral of motion

$$\dot{y} + \omega_0 x = \text{const}$$

follows. Assuming const $= 0$ without losing the generality, we arrive at equation (5.2.3) (Note 5.2).

Let us demonstrate how in equations (5.2.3) or (5.2.4) the web emerges, if the resonance condition (5.2.7) is satisfied. Let us subsequently perform a number of changes of variables. At first, let us introduce in the usual way the polar coordinates:

$$x = \rho \sin \varphi; \qquad \dot{x} = \omega_0 \rho \cos \varphi \qquad (5.2.10)$$

and make use of the expansion

$$\cos(kx - \nu t) = \cos(k\rho \sin \varphi - \nu t) = \sum_m J_m(k\rho) \cos(m\varphi - \nu t) \quad (5.2.11)$$

where J_m are the Bessel functions. Taking the advantage of the above analogy with motion in a magnetic field, ρ is the Larmor radius. The Hamiltonian (5.2.4) expressed in the new variables (5.2.10), with regard to expansion (5.2.11), acquires the following form:

$$H = \tfrac{1}{2}\omega_0^2 \rho^2 + \frac{1}{k^2} \varepsilon \omega_0^2 \sum_m J_m(k\rho) \cos(m\varphi - \nu t). \qquad (5.2.12)$$

In the sum, we single out a term with $m = n_0$

$$H = \tfrac{1}{2}\omega_0^2 \rho^2 + \frac{1}{k^2} \varepsilon \omega_0^2 J_{n_0}(k\rho) \cos(n_0\varphi - \nu t)$$

$$+ \frac{1}{k^2} \varepsilon \omega_0^2 \sum_{m \neq n_0} J_m(k\rho) \cos(m\varphi - \nu t). \qquad (5.2.13)$$

Now we introduce new variables

$$I = \omega_0 \rho^2 / 2n_0; \qquad \theta = n_0 \varphi - \nu t \qquad (5.2.14)$$

and write down the following expression:

$$\tilde{H} = H - \nu I, \qquad (5.2.15)$$

where H is expressed as a function of (I, θ). By direct calculations we can easily make sure that the equations

$$\dot{I} = -\frac{\partial \tilde{H}}{\partial \theta}; \qquad \dot{\theta} = \frac{\partial \tilde{H}}{\partial I} \qquad (5.2.16)$$

are equivalent to the equation of motion (5.2.3). By substituting (5.2.14) into (5.2.13) and (5.2.15) we get:

$$\tilde{H} = (n_0\omega_0 - \nu)I + \frac{1}{k^2}\,\varepsilon\omega_0^2 J_{n_0}(k\rho)\cos\theta$$

$$+\frac{1}{k^2}\,\varepsilon\omega_0^2 \sum_{m\neq n_0} J_m(k\rho)\cos\left[\frac{m}{n_0}\theta - \left(1 - \frac{m}{n_0}\right)\nu t\right] \qquad (5.2.17)$$

where, in search of a more compact notation, we introduce ρ which, according to (5.2.14), is as follows

$$\rho = (2n_0 I/\omega_0)^{1/2}. \qquad (5.2.18)$$

Thus, the expression $\tilde{H} = \tilde{H}(I, \theta; t)$ is the Hamiltonian with respect to the new canonical variables (I, θ). It can also be written as:

$$\tilde{H} = \tilde{H}_0(I) + \tilde{V}(I, \theta; t) \qquad (5.2.19)$$

where, in accordance with (5.2.17), we have denoted:

$$\tilde{H}_0(I) = (n_0\omega_0 - \nu)I + \frac{1}{k^2}\,\varepsilon\omega_0^2 J_{n_0}(k\rho)\cos\theta;$$

$$\qquad (5.2.20)$$

$$\tilde{V}(I, \theta; t) = \frac{1}{k^2}\,\varepsilon\omega_0^2 \sum_{m\neq n_0} J_m(k\rho)\cos\left[\frac{m}{n_0}\theta - \left(1 - \frac{m}{n_0}\right)\nu t\right]$$

while for ρ one should turn to expression (5.2.18).

Now we turn our attention to the resonance case (5.2.7). Let this condition be valid for a certain $n = n_0$, i.e., the following equality takes place:

$$n_0\omega_0 = \nu. \qquad (5.2.21)$$

This is the definition of the number n_0 which was formally introduced before. Under the condition (5.2.21), the expression for $\tilde{H}_0$ acquires the following form:

$$\tilde{H}_0 = \frac{1}{k^2}\,\varepsilon\omega_0^2 J_{n_0}(k\rho)\cos\theta = \frac{1}{k^2}\,\varepsilon\omega_0^2 J_{n_0}\left[k\left(\frac{2n_0 I}{\omega_0}\right)^{1/2}\right]\cos\theta. \qquad (5.2.22)$$

Let us perform a preliminary analysis of the dynamic system emerging in a resonance case. Both terms in the Hamiltonian, $\tilde{H}_0$ as well as $\tilde{V}$, are proportional to ε. Therefore, the stationary part of the Hamiltonian, which is time-independent, is induced by a perturbation. It disappears at $\varepsilon \to 0$. This is a new element, which is absent in the problems discussed above.

Let us take $\tilde{H}_0(I, \theta)$ for an unperturbed part of the Hamiltonian $\tilde{H}$, and $\tilde{V}(I, \theta, t)$ for a perturbation. Further, we shall see why it is possible. The phase portrait for the Hamiltonian (5.2.22) is presented in Fig. 5.2.1. Separatrices form a net on the plane $(x, \dot{x})$ in the form of a web with a number of rays n_0 and rotational symmetry by an angle $\alpha = 2\pi/n_0$ for even n_0, and with the number of rays $2n_0$ and rotational symmetry by the angle $\alpha = \pi/n_0$ for odd n_0. Singular points of a system can be found from the following equations:

$$\frac{\partial \tilde{H}_0}{\partial I} = 0; \qquad \frac{\partial \tilde{H}_0}{\partial \theta} = 0. \qquad (5.2.23)$$

Hence, if we substitute expression (5.2.22), for $\tilde{H}_0$, there follows a set of hyperbolic points (ρ_h, θ_h)

$$J_n(k\rho_h) = 0; \qquad \theta_h = \pm\pi/2 \qquad (5.2.24)$$

and elliptic points (ρ_e, θ_e)

$$J'_{n_0}(k\rho_e) = 0; \qquad \theta_e = 0, \pi. \qquad (5.2.25)$$

A family of separatrices is formed by $2n_0$ rays and by circumferences with the radii $\rho_h^{(s)}$, crossing them, where $k\rho_h^{(s)}$ are radicals of the Bessel function J_{n_0}. Inside the cells of a web formed by these separatrices, motion occurs along closed-type orbits, round the elliptic points. These

Fig. 5.2.1   The phase portrait of a system with a separatrix network in the form of a web: $n_0 = 4$.

p

x

points (5.2.25) lie in the centres of cells. The above is the description of unperturbed motion defined by the Hamiltonian (5.2.22). There is a principal difference between this motion and the motion, say, of a nonlinear pendulum. A particle travelling along a web can move in a radial direction, while for systems of the pendulum type, any motion in a radial direction is ruled out. This is a principal characteristic of the nonlinear case, for which (5.1.6) holds true.

However, radial motion which, on the average, is non-zero, is possible only along separatrices. In the vicinity of hyperbolic points it is frozen. Therefore, there is no progress along the radius for the Hamiltonian $\tilde{H}_0$ (5.2.22). If we take into account the effect of perturbation $\tilde{V}$ (5.2.20) this should result in the destruction of separatrices and the formation of channels of a finite width, with stochastic dynamics within them, in their place (Fig. 5.2.2).

5.3 Invariant tori inside the web (web-tori) and the width of the web

Now we shall discuss the motion within the web in more detail. Let us go back to the original Hamiltonian (5.2.19) in order to somewhat simplify it. We assume that the resonance condition (5.2.21) holds true and retain in (5.2.20) for $\tilde{V}$ only the terms with $m = n_0 \pm 1$. As will be shown later, the contribution of other terms in the sum for $\tilde{V}$ is smaller (a similar technique was already applied in computation of the width of the stochastic layer in Sect. 3.4). Finally, (5.2.20) yields:

$$\tilde{H} = \tilde{H}_0 + \tilde{V}; \qquad \tilde{H}_0 = \frac{1}{k^2}\,\varepsilon\omega_0^2 J_{n_0}(k\rho)\cos\theta;$$

$$\tilde{V} = \frac{1}{k^2}\,\varepsilon\omega_0^2 \left\{ J_{n_0+1}(k\rho)\cos\left[\left(1+\frac{1}{n_0}\right)\theta + \frac{1}{n_0}\,\nu t\right] \right. \tag{5.3.1}$$
$$\left. + J_{n_0-1}(k\rho)\cos\left[\left(1-\frac{1}{n_0}\right)\theta - \frac{1}{n_0}\,\nu t\right] \right\}.$$

Fig. 5.2.2 The trajectory indicated by the arrows in (a) is impossible in reality since it would take an infinitely long time to reach a hyperbolic point. However, it is possible in the case of (b) when a stochastic layer is formed.

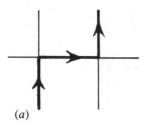
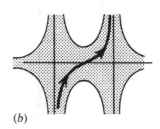

(a) (b)

For simplicity, let us also consider regions sufficiently remote from the centre of a web, i.e., we put:

$$k\rho \gg 1. \qquad (5.3.2)$$

Let us make use of the asymptotics of the Bessel functions, in accordance with (5.3.2):

$$J_n(k\rho) \sim \left(\frac{2}{\pi k\rho}\right)^{1/2} \cos(k\rho - \tfrac{1}{2}\pi n - \tfrac{1}{4}\pi). \qquad (5.3.3)$$

Now, we single out a certain cell of the web and describe the family of trajectories inside it, at first omitting perturbation $\tilde{V}$ (Fig. 5.3.1). Let ρ_0 be an elliptic point in the centre of the cell. According to (5.2.25) and (5.3.3), in the case of (5.3.2), we have

$$k\rho_0 - \tfrac{1}{2}\pi n_0 - \tfrac{1}{4}\pi = 0; \; \pi. \qquad (5.3.4)$$

With the help of expansion (5.3.3) and condition (5.3.4) we rewrite (5.3.1) as follows:

$$\tilde{H}_0 = -\sigma\varepsilon \frac{\omega_0^2}{k^2}\left(\frac{2}{\pi k\rho_0}\right)^{1/2} \cos k\tilde{\rho} \cos\theta, \qquad \sigma = \pm 1, \qquad (5.3.5)$$

where $\tilde{\rho} = \rho - \rho_0$ and different signs correspond to different coordinates of elliptic points, depending on the value of the right-hand side of (5.3.4).

Let us dwell on the analysis of trajectories determined by the Hamiltonian (5.3.5). The size of a separatrix cell is of the order of $2\pi/k$.

Fig. 5.3.1 One cell of a web with an elliptic point ρ_0 in the centre and a family of orbits inside.

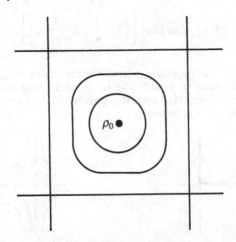

Therefore, max $\Delta \rho = 2\pi / k$ and, according to inequality (5.3.2):

$$|\tilde{\rho}| = |\rho - \rho_0| \ll \rho_0.$$ (5.3.6)

The definition of action (5.2.14) yields

$$I = \frac{1}{2n_0} \, \omega_0 \rho^2 \approx \frac{1}{2n_0} \, \omega_0 \rho_0^2 + \frac{\omega_0 \rho_0}{n_0} \, \tilde{\rho}.$$ (5.3.7)

The following change of a variable

$$I = I - \frac{1}{2n_0} \, \omega_0 \rho_0^2$$

is canonical (shift by a constant). With the same approximation as for the Hamiltonian $\tilde{H}$, the pair of variables $(\tilde{\rho}, \theta)$ may be considered canonical. Accordingly, the Hamiltonian equations of motion have the following form:

$$\dot{\tilde{\rho}} = -\frac{n_0}{\rho_0 \omega_0} \frac{\partial \tilde{H}_0}{\partial \theta}; \qquad \dot{\theta} = \frac{n_0}{\rho_0 \omega_0} \frac{\partial \tilde{H}_0}{\partial \tilde{\rho}}.$$ (5.3.8)

Indeed, they coincide with the ones following from (5.2.16) under condition (5.3.6).

Let us denote

$$\omega_{\mathrm{w}} = -\sigma \left(\frac{2}{\pi} \right)^{1/2} \frac{\varepsilon n_0 \omega_0}{(k\rho_0)^{3/2}}$$ (5.3.9)

and assume:

$$H_{\mathrm{w}} = \omega_{\mathrm{w}} \cos \xi \cos \theta$$ (5.3.10)

where $\xi = k\tilde{\rho}$. Then equations

$$\dot{\xi} = \frac{\partial H_{\mathrm{w}}}{\partial \theta}; \qquad \dot{\theta} = -\frac{\partial H_{\mathrm{w}}}{\partial \xi}$$ (5.3.11)

are equivalent to (5.3.8). The Hamiltonian H_{w} can be called the Hamiltonian of the web-tori. ω_{w} is the frequency of small oscillations for trajectories wound around web-tori.

Let us rewrite (5.3.11) in an explicit form:

$$\dot{\xi} = -\omega_{\mathrm{w}} \cos \xi \sin \theta; \qquad \dot{\theta} = \omega_{\mathrm{w}} \sin \xi \cos \theta.$$ (5.3.12)

By making use of (5.3.10) and integrating (5.3.12), we get the following:

$$\sin \theta = \varkappa \, \mathrm{sn}(\omega_{\mathrm{w}}(t - t_0); \varkappa)$$ (5.3.13)

where sn is the Jacobian elliptic function, t_0 is the moment of time at which $\theta = 0$ and $\varkappa$ is the modulus of the elliptic function:

$$\varkappa = (1 - H_w^2/\omega_w^2)^{1/2}. \tag{5.3.14}$$

By means of solution (5.3.13) and expressions (5.3.10) and (5.3.14), we get the following:

$$\sin \xi = \varkappa \frac{\text{cn}(\omega_w(t - t_0); \varkappa)}{\text{dn}(\omega_w(t - t_0); \varkappa)}. \tag{5.3.15}$$

The solutions (5.3.13) and (5.3.15) are periodic functions of time. The period of nonlinear oscillations is:

$$T(H_w) = \frac{1}{|\omega_w|} 4K(\varkappa) \tag{5.3.16}$$

where $K(\varkappa)$ is a full elliptic integral of the first kind. For $\varkappa \to 0$ we have:

$$T_w = 2\pi/|\omega_w|$$

i.e., the period of small oscillations. Near a separatrix, $\varkappa \to 1$ and from equation (5.3.16) it follows that

$$T(H_w) = \frac{1}{|\omega_w|} 4 \ln \frac{4}{(1-\varkappa^2)^{1/2}} = \frac{1}{|\omega_w|} 4 \ln \frac{4|\omega_w|}{|H_w|}$$

$$= \left(\frac{\pi}{2}\right)^{1/2} \frac{4}{\varepsilon n_0 \omega_0} (k\rho_0)^{3/2} \ln\left[4\varepsilon \frac{\omega_0^2}{k^2}\left(\frac{2}{\pi k\rho_0}\right)^{1/2} \frac{1}{\tilde{H}_0}\right] \tag{5.3.17}$$

On separatrices, $H_w = 0$. From (5.3.10), it follows that four separatrices are defined by the following equations:

$$\cos \xi = 0; \qquad \xi = \pm\pi/2 \tag{5.3.18}$$

and

$$\cos \theta = 0; \qquad \theta = \pm\pi/2. \tag{5.3.19}$$

They correspond to the four sides of the square (under certain approximation, $\tilde{\rho} \ll \rho_0$) in Fig. 5.3.1. The following equation of motion:

$$\sin \theta = \pm\tanh[2|\omega_w|(t - t_0)]; \qquad \xi = \pm\pi/2 \tag{5.3.20}$$

corresponds to two horizontal separatrices in (5.3.13) and (5.3.18). For two vertical separatrices in (5.3.15) and (5.3.19), we obtain:

$$\sin \xi = \mp\tanh[2|\omega_w|(t - t_0)], \qquad \theta = \mp\pi/2 \tag{5.3.21}$$

(Fig. 5.3.2) (Note 5.3).

The closed trajectories defined by the Hamiltonian (5.3.5) and (5.3.10) are the cross-sections of invariant tori if we complement the phase space (I, θ) with the variable 'time' in the usual way, by taking into account the perturbation $\tilde{V}$, which is periodic in time. The invariant tori lying inside a web we shall call web-tori. Their difference from KAM-tori manifests itself in the dependency of period T_w on ε. In the case of KAM-tori, $T \sim 1/\varepsilon^{1/2}$, while in the case of web-tori, $T_w \sim 1/\varepsilon$.

Now let us turn our attention to the picture of destruction of separatrices and formation of a stochastic web in their place. With this aim in view, we consider the original Hamiltonian (5.3.1). Making use of the asymptotics (5.3.3), we can rewrite the perturbation $\tilde{V}$ in the following way:

$$\tilde{V} \approx 2\varepsilon \frac{\omega_0^2}{k^2} \left(\frac{2}{\pi k \rho_0} \right)^{1/2} \sigma \sin k\tilde{\rho} \sin \theta \sin \left[\frac{1}{n_0} (\theta + \nu t) \right]. \quad (5.3.22)$$

Thus, the whole problem, according to (5.3.5) and (5.3.22), is described by the following Hamiltonian:

$$\tilde{H} = \sigma \varepsilon \frac{\omega_0^2}{k^2} \left(\frac{2}{\pi k \rho_0} \right)^{1/2} \left\{ -\cos k\tilde{\rho} \cos \theta \right.$$

$$\left. + 2 \sin k\tilde{\rho} \sin \theta \sin \left[\frac{1}{n_0} (\theta + \nu t) \right] \right\}. \quad (5.3.23)$$

Fig. 5.3.2   Various separatrix branches of a web's cell.

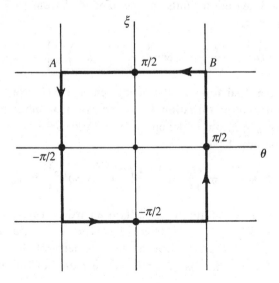

Hence

$$\overset{\star}{\tilde{H}}_0 = \frac{\partial \tilde{H}_0}{\partial \theta}\dot{\theta} + \frac{\partial \tilde{H}_0}{\partial \tilde{\rho}}\overset{\star}{\rho} = \frac{n_0}{\rho_0 \omega_0}\left(\frac{\partial \tilde{H}_0}{\partial \theta}\frac{\partial \tilde{V}}{\partial \tilde{\rho}} - \frac{\partial \tilde{H}_0}{\partial \tilde{\rho}}\frac{\partial V}{\partial \theta}\right)$$

$$= \frac{1}{\pi}2n_0\varepsilon^2 \frac{\omega_0^2}{k^2(k\rho_0)^2}\sin\left(\frac{1}{n_0}\theta + \omega_0 t\right)(\cos 2k\tilde{\rho} - \cos 2\theta).$$

$$(5.3.24)$$

On a stretch of the trajectory in the vicinity of a separatrix AB (Fig. 5.3.2) we have:

$$k\tilde{\rho} = \pi/2; \qquad \cos 2k\tilde{\rho} = -1.$$

Therefore,

$$\overset{\star}{\tilde{H}}_0 = -\frac{4n_0\varepsilon^2\omega_0^3}{\pi k^2(k\rho_0)^2}\cos^2\theta\sin\omega_0 t, \qquad n_0 \gg 1. \qquad (5.3.25)$$

Let us compute the change near a separatrix $\Delta\tilde{H}_0$ under the effect of perturbation. Integration of (5.3.25) yields

$$\Delta\tilde{H}_0 = -\frac{4n_0\varepsilon^2\omega_0^3}{\pi k^2(k\rho_0)^2}\int dt\frac{\sin\omega_0 t}{\cos^2\theta}.$$

We can substitute into this formula the value of $\cos\theta$ taken on a separatrix. From (5.3.20), we have:

$$\cos\theta = 1/\cosh[2|\omega_w|(t - t_0)].$$

Besides, the integration limits can be expanded from $t = -\infty$ to $t = \infty$. As a result, we have

$$\Delta\tilde{H}_0 = -\frac{4\pi\omega_0^2}{n_0 k^2}k\rho_0 \exp\left\{-\frac{1}{\varepsilon n_0}\left(\frac{\pi}{2}k\rho_0\right)^{3/2}\right\}\sin\omega_0 t_0. \qquad (5.3.26)$$

To obtain the final form of the energy change $\Delta\tilde{H}_0$ near a separatrix under the effect of perturbation, let us express the number of resonances n_0 through $k\rho_0$ by means of definition (5.3.4) and condition $n_0 \gg 1$. Instead of (5.3.4) we now have

$$\Delta\tilde{H}_0 = -2\pi^2\frac{\omega_0^2}{k^2}\exp\left\{-\frac{1}{\varepsilon}\left(\frac{\pi}{2}\right)^{5/2}(k\rho_0)^{1/2}\right\}\sin\omega_0 t_0. \qquad (5.3.27)$$

We have considered the change $\Delta\tilde{H}_0$ as a particle travels in the vicinity of one of the separatrices, AB (Fig. 5.3.2). The same is true for the three other branches of the cell. Due to the symmetry of the problem, the expression for $\Delta\tilde{H}_0$ is alike for all four segments of the orbit, so that only the moments of time t_0 vary, which define the position of a particle as it passes through the middle of one of the four segments of the orbit in the vicinity of a corresponding separatrix. The time lapse between two subsequent traversals of the middle points of the separatrices is equal to one quarter of a period, i.e., $I(H_w)/4$. This enables us to describe the dynamics of a system inside a cell, corresponding to the Hamiltonian $\tilde{H}$ (5.2.23), as a mapping for the variables $\tilde{H}_0$ and $\psi = \omega(H_w)t$, where

$$\omega(H_w) = 2\pi/T(H_w) \tag{5.3.28}$$

is the frequency of nonlinear oscillations inside a separatrix cell.

By making use of the formulas (5.3.27) and (5.3.17), we get the following mapping near separatrices of a web:

$$\tilde{H}_{n+1} = \tilde{H}_n - 2\pi^2 \frac{\omega_0^2}{k^2} \exp\left\{-\frac{1}{\varepsilon}\left(\frac{\pi}{2}\right)^{5/2} (k\rho_0)^{1/2}\right\} \sin \psi_n,$$

$$\tag{5.3.29}$$

$$\psi_{n+1} = \psi_n - \frac{1}{\varepsilon}\left(\frac{\pi}{2}\right)^{3/2} (k\rho_0)^{1/2} \ln\left[4\varepsilon \frac{\omega_0^2}{k^2}\left(\frac{2}{\pi k\rho_0}\right)^{1/2} \frac{1}{\tilde{H}_{n+1}}\right].$$

The following condition of stochastic dynamics

$$K \equiv |\partial\psi_{n+1}/\partial\psi_n - 1| \geqslant 1$$

defines H_c, the border of a stochastic layer forming in the neighbourhood of separatrices. From (5.3.29), for $K = 1$, we have the following:

$$H_c = 2^{-1/2}\pi^{7/2} \frac{(k\rho_0)^{1/2}\omega_0^2}{\varepsilon k^2} \exp\left\{-\frac{1}{\varepsilon}\left(\frac{\pi}{2}\right)^{5/2} (k\rho_0)^{1/2}\right\}. \tag{5.3.30}$$

The width of the stochastic web is $2H_c$. As follows from (5.3.30), it is exponentially small in parameter ε and, besides, falls off exponentially as ρ_0 increases, i.e., the greater the distance from the centre of the web. A numerical example of a stochastic web and the distribution function for it is shown in Fig. 5.3.3 [10].

Thus, different web-tori are separated by a stochastic web. It forms a uniform network of an exponentially small width. In the exponent we see the factor of $1/\varepsilon$ in contrast to $1/\varepsilon^{1/2}$, for the Arnold web. Besides, the thinning of the web as the energy of a particle grows, i.e., as the particle moves further away from the centre of the web, impedes the diffusion along the web over considerable distances and practically cuts it off.

Fig. 5.3.3   The stochastic web (*a*) on the phase plane of equation
(5.2.3) at $\varepsilon\omega_0^2/k = 0.1$, $k = 15$, $\nu = 4\omega_0$, $\omega_0 = 1$ and the distribution
function of the phase points in it (*b*), as the result of 2×10^4 periods
of oscillation $2\pi/\omega_0$.

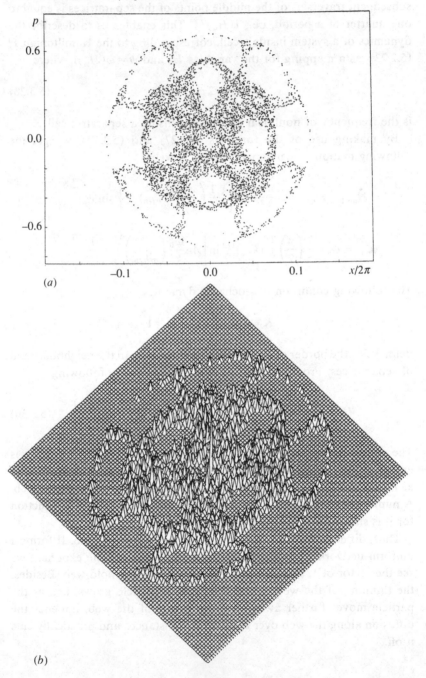

(*a*)

(*b*)

5.4 The KAM-tori to web-tori transition

We have shown above that for the following equation

$$\ddot{x} + \omega_0^2 x = \frac{1}{k}\,\varepsilon\omega_0^2 \sin(kx - \nu t), \tag{5.4.1}$$

in the case of $\varepsilon \ll 1$ and under a resonance condition

$$\nu = n_0\omega_0, \tag{5.4.2}$$

the phase portrait looks like a set of invariant web-tori inside the cells of a web. The web itself is stochastic, i.e., it has a finite width, and inside it the dynamics of a particle is stochastic. The web exists for arbitrarily small ε. A strong degeneracy in the problem can be removed in two ways: by detuning of resonance (5.4.1) or by an introduction of non-linearity into the left-hand side of equation (5.4.1).

In the present section of the book we shall see what happens when both kinds of degeneracy (due to resonance and due to the linearity of the unperturbed problem) are removed, although the system still remains close to a degenerate state (Note 5.4).

Let us once again consider the system with the Hamiltonian

$$H = \tfrac{1}{2}\dot{x}^2 - \omega_0^2 \cos x + \frac{1}{k^2}\,\varepsilon\omega_0^2 \cos(kx - \nu t) \tag{5.4.3}$$

which describes the motion of a particle in the field of two plane waves. It also corresponds to a nonlinear pendulum affected by a plane wave ($\varepsilon \ll 1$).

Let us assume that the perturbation in (5.4.3) is small-scale, i.e.,

$$k \gg 1. \tag{5.4.4}$$

Condition (5.4.3) is of paramount importance and later we shall see why. The equations of motion following from (5.4.3) have the following form:

$$\ddot{x} + \omega_0^2 \sin x = \frac{1}{k}\,\varepsilon\omega_0^2 \sin(kx - \nu t). \tag{5.4.5}$$

Now we expand $\cos x$ in (5.4.3) up to the third term and $\sin x$ in (5.4.5) up to the second term:

$$\begin{aligned}
\sin x &\approx x - x^3/6 \\
\cos x &\approx 1 - x^2/2 + x^4/24.
\end{aligned} \tag{5.4.6}$$

When $x \ll 1$, the contribution of nonlinear corrections x^4, or x^3, is small.

Once again we introduce the polar coordinates (ρ, φ), by means of equations (5.2.10):

$$x = \rho \sin \varphi; \qquad \dot{x} = \omega_0 \rho \cos \varphi. \qquad (5.4.7)$$

Assuming, as in (5.2.14),

$$\theta = n_0 \varphi - \nu t$$

we get

$$x = \rho \sin \left[\frac{1}{n_0} (\theta + \nu t) \right];$$

$$\qquad (5.4.8)$$

$$\dot{x} = \omega_0 \rho \cos \left[\frac{1}{n_0} (\theta + \nu t) \right].$$

To avoid the cumbersome presentation, let us briefly outline the process of transfer. For the main part of the Hamiltonian (the pendulum) we can introduce the action-angle variables. This has already been done. However, our major interest is in the region of small values of ρ. Therefore, this part of the Hamiltonian can be obtained by simple averaging, after the expansions (5.4.6) and (5.4.7) have been substituted into (5.4.3). This yields

$$H_\rho = \omega_0 I + aI^2, \qquad a = -n_0^2/16 \qquad (5.4.9)$$

where

$$I = \omega_0 \rho^2 / 2 n_0 \qquad (5.4.10)$$

just as in (5.2.14).

Now we can use the old notation for the Hamiltonian (5.2.20), having added only one nonlinear term:

$$H = H_0(I, \theta) + V(I, \theta; t)$$

$$H_0(I, \theta) = (n_0 \omega_0 - \nu) I + aI^2 + \frac{1}{k^2} \varepsilon \omega_0^2 J_{n_0}(k\rho) \cos \theta \qquad (5.4.11)$$

$$V(I, \theta; t) = \frac{1}{k^2} \varepsilon \omega_0^2 \sum_{m \neq n_0} J_m(k\rho) \cos \left[\frac{m}{n_0} \theta - \left(1 - \frac{m}{n_0} \right) \nu t \right]$$

where it is assumed that ρ is expressed through action I, according to equation (5.4.10). The structure of equations (5.4.11) is essentially the same as of (5.3.1). Only two terms in $H_0(I, \theta)$ have been added. The first is caused by detuning of resonance, while the second owes its existence to the seed nonlinearity. When $\varepsilon = 0$, the frequency is defined by the following expression:

$$\delta\omega \equiv \frac{\partial H_0(I, \theta; \varepsilon = 0)}{\partial I} = n_0\omega_0 - \nu + 2aI \equiv \delta\omega_L + \delta\omega_N \qquad (5.4.12)$$

where $\delta\omega_L = n_0\omega_0 - \nu$ and $\delta\omega_N = 2aI$ and are linear and nonlinear parts respectively. Since $\delta\omega \neq 0$ the phase portrait of a system in the case of $V = 0$ already has no web (Fig. 5.4.1): degeneracy is taken off. Now in phase space there appear invariant curves embracing the centre. They impede diffusion in a radial direction.

However, although the situation is similar to the one observed in the case of the KAM theory, many of the curves, sharply twisting, form a picture that closely resembles a web. The further reasoning is simple enough: if a perturbation is sufficient, so that the resulting stochastic layer is wide enough to cover the gaps between separatrices, then a single large web can appear. This statement is illustrated by Fig. 5.4.2 where a typical separatrix cell for a perturbed pendulum is depicted. Under the effect of a perturbation, a separatrix is destroyed and an extremely narrow stochastic layer is formed in its vicinity. Further, within the cell lie the curves which are cross-sections of KAM-tori. Closer to the cell's centre there lies a region which resembles the cross-section of a web-torus (Fig. 5.4.2b).

Two additional examples in Fig. 5.4.3 show how, due to the resonance condition (5.4.2) and condition (5.4.4), a finite portion of the web can get implanted in a separatrix. The greater the value of k, the smaller the

Fig. 5.4.1 The lines of the level of the averaged Hamiltonian, $H_0(I, \theta) = \text{const}$, define a family of particles' trajectories $n_0 = 4$: (a) $\delta\omega_N = 0$, $(k^2/\varepsilon\omega_0^2)\delta\omega_L = 0.03$; (b) $\delta\omega_L = 0$, $(k^2 a/\varepsilon\omega_0^2) = 0.001$.

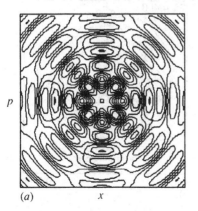

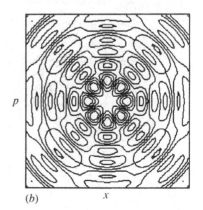

p

p

(a)　　　x　　　(b)　　　x

cells of the web, since their size is of the order of magnitude $2\pi/k$. The greater the value of ε, the wider the stochastic web. If the parameters k and ε are not large enough, the picture of the inside of a basic separatrix cell has various intermediate states (Fig. 5.4.4). These pictures show how,

Fig. 5.4.2 The implantation of a part of a web in a standard separatrix cell in the problem of a distorted pendulum, equation (5.4.5): $\omega_0 = 1$, $\nu = 4$, $k = 75$. (*a*) phase portrait inside the separatrix cell; (*b*) enlargement of the internal region shown in (*a*).

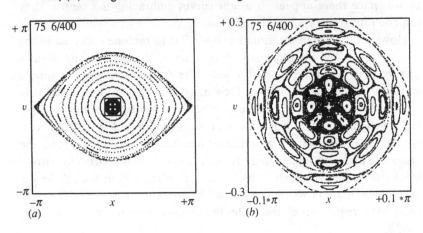

Fig. 5.4.3 Two examples of the internal region inside a separatrix cell: (*a*) $k = 300$, $\varepsilon/k = 1/400$, $n_0 = 4$, $\omega_0 = 1$; (*b*) $k = 75$, $\varepsilon/k = 1/100$, $n_0 = 4$, $\omega_0 = 1$.

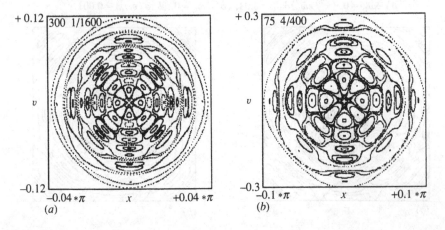

in the process of deformation of the cross-sections of KAM-tori, new smaller cells of web-tori are formed as parameter k increases.

The above changes are called the KAM-tori to web-tori transition. During this transition, as ε increases, there occur an infinite number of

Fig. 5.4.4   Various studies of the KAM-tori to web-tori transition: $\omega_0 = 1$, $\nu = 5$, $k = 12$. (a) a complete separatrix cell at $\varepsilon/k = 1/80$; (b) the internal part of a portrait in (a); (c) as (b) but at $\varepsilon/k = 1/30$; (d) as (b) but at $\varepsilon/k = 1/20$.

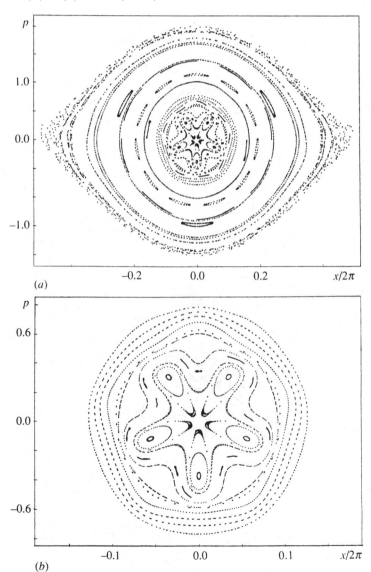

bifurcational changes of the phase portrait. The patterns on the phase plane which owe their existence to a closeness to degeneracy, are supplementary to the basic picture of the phase portrait following from the KAM theory.

Fig. 5.4.4—continued

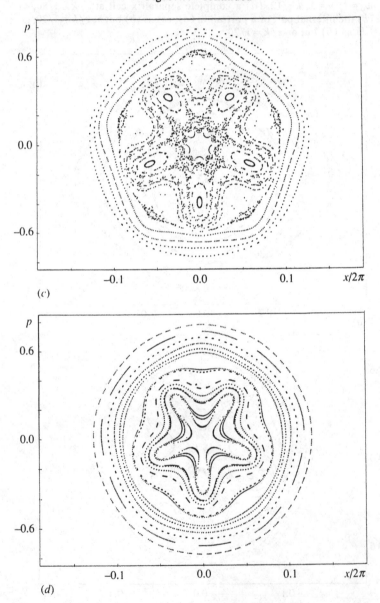

(c)

(d)

6

The uniform web

The web discussed in the previous chapter was infinite. However, the width of this web was diminishing exponentially as we moved away from its centre, making the web effectively finite. In this chapter we are going to discuss the case when the width of a web, on average, does not change in the whole phase plane. This web we shall call a uniform one. It first appeared in an analysis of the motion of charged particles in a magnetic field and a field of a perpendicular wave packet with a large number of harmonics [1, 2]. Later, the problem turned out to be of much greater interest, since it is related to the problem of symmetry in plane tilings, to random walks along patterns, to hydrodynamic patterns and to the geometry of quasi-crystals (see [3] and [4]).

In thic chapter we shall introduce a new type of mapping, conserving the measure, to which the KAM theory cannot be directly applied. We obtain, thereby, a broader understanding of the nature of chaos in Hamiltonian systems.

6.1 Mapping with a twist

Initial equations of a particle's motion have the following form

$$\ddot{\boldsymbol{r}} = \frac{e}{m} \boldsymbol{E}(x, t) + \frac{e}{mc} [\dot{\boldsymbol{r}}, \boldsymbol{B}_0] \qquad (6.1.1)$$

where $\boldsymbol{r} = (x, y)$ is a vector lying in the plane (x, y); $\boldsymbol{E}$ is a wave packet propagating along the x-axis and directed along x, and $\boldsymbol{B}_0$ is a constant magnetic field parallel to the z-axis. The absence of a y-dependency on the right-hand side of the equation leads to an integral of motion

$$\dot{y} = v_y = -\omega_0 x + \text{const}$$

109

where

$$\omega_0 = eB_0/mc$$

is the cyclotron frequency. Without neglecting the generality, we may set const $= 0$. Then

$$\dot{y} = -\omega_0 x. \tag{6.1.2}$$

For a wave packet $E(x, t)$ we assume the following expression

$$E = -E_0 \sum_{n=-\infty}^{\infty} \sin(kx - \omega t - n\nu t)$$

$$= -E_0 T \sin(kx - \omega t) \sum_{n=-\infty}^{\infty} \delta(t - nT), \tag{6.1.3}$$

where E_0 is directed along x and

$$T = 2\pi/\nu. \tag{6.1.4}$$

By substituting (6.1.2) and (6.1.3) into (6.1.1), we get

$$\ddot{x} + \omega_0^2 x = -\frac{e}{m} E_0 T \sin(kx - \omega t) \sum_{n=-\infty}^{\infty} \delta(t - nT) \tag{6.1.5}$$

– an equation describing a linear oscillator affected by short (δ-like) kicks with the period T. To transfer from (6.1.5) to a finite difference equation, we denote

$$x_n = x(t_n - 0); \qquad \dot{x}_n = v_{x,(n)} = \dot{x}(t_n - 0).$$

The relation between the values of x and $\dot{x}$ to the left and to the right of the δ-function, has the following form

$$x(t_n + 0) = x(t_n - 0),$$

$$\dot{x}(t_n + 0) = \dot{x}(t_n - 0) - \frac{e}{m} TE_0 \sin(kx_n - \omega t_n),$$

where $t_n = nT$. These relations lead to the following mapping

$$v_{x,(n+1)} = \omega_0 x_n \sin \omega_0 T + \left[v_{x,(n)} + \frac{e}{m} E_0 T \sin(n\omega T - kx_n) \right] \cos \omega_0 T,$$

$$x_{n+1} = x_n \cos \omega_0 T + \frac{1}{\omega_0} \left[v_{x,(n)} + \frac{e}{m} E_0 T \sin(n\omega T - kx_n) \right] \sin \omega_0 T. \tag{6.1.6}$$

For $\omega_0 \to 0$, the mapping (6.1.6) is reduced to the standard mapping

$$v_{x,(n+1)} = v_{x,(n)} + \frac{e}{m} E_0 T \sin kx_n,$$

$$x_{n+1} = x_n + T(v_{x,(n+1)} - \omega/k) \tag{6.1.7}$$

already encountered in Sect. 3.4. In the case of (6.1.7), global chaos appears on the following condition

$$K = \frac{e}{m} E_0 k T^2 \gtrsim 1. \qquad (6.1.8)$$

The phase portrait of the system (6.1.7) for $K \ll 1$, is also described in Sect. 3.4. Narrow stochastic layers are separated by invariant curves, which make any advancement in the direction perpendicular to the layers impossible (i.e., an increase of energy is ruled out). As will be shown later, the properties of the mapping (6.1.6) are completely different.

The term $\omega T \neq 2\pi m$ (where m is an integer) in the sine's argument in (6.1.6) has a simple physical meaning. One only has to make the following change of variables

$$\tilde{x} = x - n\omega T / k.$$

It follows from equation (6.1.2) that the particles are accelerating. The change of velocity with one kick is $\omega \omega_0 T / k$. Therefore, over time t, velocity changes by a value

$$\Delta v \sim \omega_0 t (\omega / k) \qquad (6.1.9)$$

where ω / k is the phase velocity of a wave packet. Acceleration is due to a wave regularly pushing the particle (Note 6.1).

Further we set $\omega = 0$, thus eliminating any possibility of a regular acceleration of the type of (6.1.9). We denote

$$\alpha = \omega_0 T. \qquad (6.1.10)$$

Let us introduce new dimensionless variables

$$u = k v_x / \omega_0, \qquad v = k v_y / \omega_0 = -kx. \qquad (6.1.11)$$

Now the equations are reduced to the following mapping

$$\hat{M}_\alpha : \begin{cases} \bar{u} = (u + K_0 \sin v) \cos \alpha + v \sin \alpha \\ \bar{v} = -(u + K_0 \sin v) \sin \alpha + v \cos \alpha \end{cases} \qquad (6.1.12)$$

where the indexes n and $n+1$, are omitted for the sake of simplicity, and we denote

$$K_0 = K / \alpha. \qquad (6.1.13)$$

We shall call the mapping $\hat{M}_\alpha$ the mapping with a twist about the angle α. Let us write out the Hamiltonian of the problem resulting in equations (6.1.12). The following Hamiltonian corresponds to the equation of motion (6.1.5) at $\omega = 0$

$$H = \tfrac{1}{2}(\dot{x}^2 + \omega_0^2 x^2) - \frac{e}{m} E_0 T \cos kx \sum_{n=-\infty}^{\infty} \delta(t - nT). \qquad (6.1.14)$$

Expressed in the dimensionless variables (u, v), the Hamiltonian has the following form

$$H = \tfrac{1}{2}\alpha(u^2+v^2) - K_0 \cos v \sum_{n=-\infty}^{\infty} \delta(\tau-n) \qquad (6.1.15)$$

where we introduce the dimensionless time

$$\tau = t/T.$$

The equations of motion have the following form

$$\frac{du}{d\tau} = \frac{\partial H}{\partial v} = \alpha v + K_0 \sin v \sum_{n=-\infty}^{\infty} \delta(\tau-n)$$
$$\frac{dv}{dt} = -\frac{\partial H}{\partial u} = -\alpha u. \qquad (6.1.16)$$

Expressed in the variables (u, v) and τ, the oscillator has the frequency α, while kicks follow with the time interval $\Delta\tau = 1$. We are going to pay special attention to a resonance case. A resonance occurs when an integer number of kicks q occurs during one period of oscillations of the oscillator, $2\pi/\alpha$. The equality $2\pi/\alpha_q = q$ or

$$\alpha_q = 2\pi/q \qquad (6.1.17)$$

serves as a condition for this. If we recall notation (6.1.10), the resonance condition (6.1.17) is equivalent to

$$q\omega_0 = 2\pi/T$$

and coincides with condition (5.2.7), where for the frequency of perturbation we should set $\nu = 2\pi/T$. The mapping $\hat{M}_q$, in this case, is obtained from $\hat{M}_\alpha$, if we substitute the value of (6.1.17) for α into (6.1.12)

$$\hat{M}_q: \begin{cases} \bar{u} = (u + K_0 \sin v)\cos\left(\frac{2\pi}{q}\right) + v\sin\left(\frac{2\pi}{q}\right) \\ \bar{v} = -(u + K_0 \sin v)\sin\left(\frac{2\pi}{q}\right) + v\cos\left(\frac{2\pi}{q}\right). \end{cases} \qquad (6.1.18)$$

Let us present several special cases of $\hat{M}_q$. For $q=1$

$$\hat{M}_1: \begin{cases} \bar{u} = u + K_0 \sin v \\ \bar{v} = v. \end{cases} \qquad (6.1.19)$$

Hence,

$$v = \text{const} = v_0; \qquad u_n = u_0 + K_0 n \sin v_0. \qquad (6.1.20)$$

The mapping $\hat{M}_1$ corresponds to the case of the principal cyclotron resonance $(\omega_0 = \nu)$ while the solution of equation (6.1.20) describes an acceleration along the x-axis and n plays the part of discrete time $(n = t/T)$.

For $q = 2$, we have from (6.1.18)

$$\hat{M}_2: \begin{cases} \bar{u} = -u - K_0 \sin v \\ \bar{v} = -v. \end{cases} \qquad (6.1.21)$$

This case corresponds to a half-integer cyclotron resonance. Consider two consequent mappings $\hat{M}_2\hat{M}_2 = \hat{M}_2^2$. It follows from (6.1.21) that

$$\hat{M}_2^2: \begin{cases} \bar{\bar{u}} = -\bar{u} - K_0 \sin \bar{v} = u + 2K_0 \sin v \\ \bar{\bar{v}} = -\bar{v} = v. \end{cases}$$

These expressions coincide with (6.1.19) and, therefore, have the same solution (6.1.20), describing the acceleration of a particle.

The above cases of resonance with $q = 1, 2$, as we have seen, are very simple, i.e., they admit of an accurate solution. Beginning with $q > 2$, the systems (6.1.15) and (6.1.16) are no longer integrable and their analysis, as will be shown below, is rather complex.

6.2 The periodic web

In this section we shall discuss the resonances with $q = 3, 4$ and 6. Later we shall see what makes us classify them together.

Let us assume in (6.1.18) that $q = 4$. We have

$$\hat{M}_4: \begin{cases} \bar{u} = v \\ \bar{v} = -u - K_0 \sin v. \end{cases} \qquad (6.2.1)$$

The Hamiltonian corresponding to the mapping $\hat{M}_4$ can be derived from (6.1.15)

$$H = \omega_4(u^2 + v^2) - K_0 \cos v \sum_{n=-\infty}^{\infty} \delta(\tau - n), \qquad \omega_4 = \pi/4. \qquad (6.2.2)$$

The phase portrait of the mapping $\hat{M}_4$ is shown in Fig. 6.2.1. On the phase plane, there is an infinite web, i.e., a region of stochastic dynamics of a particle. Inside a cell of the web, the family of orbits mostly consists of closed-type periodic trajectories which are cross-sections of invariant tori. For small values of K_0, the web is thin and resembles a square lattice. With an increase of K_0, the region of stochastic dynamics expands, while the size of stability islands inside the cells diminishes, although their arrangement still retains the symmetry of a square lattice (Fig. 6.2.2).

Fig. 6.2.1 The phase portrait on the plane (u, v) of the mapping $\hat{M}_4$ has a periodic web with the symmetry of a square lattice: (*a*) $K_0 = 0.7$; the size of the square $= 24\pi \times 24\pi$; (*b*) $K_0 = 1.5$; the size of the square $= 16\pi \times 16\pi$.

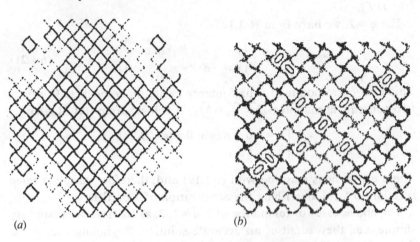

(*a*)   (*b*)

Fig. 6.2.2 The phase portrait on the plane (u, v) of the mapping $\hat{M}_4$ at a large value of $K_0 = 3$; the size of the square $= 4\pi \times 4\pi$.

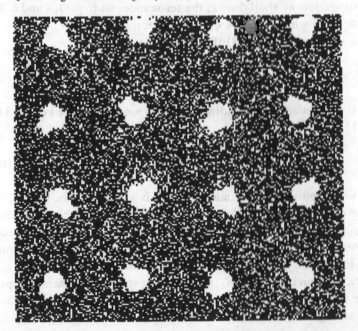

One more numerical example presented in Fig. 6.2.3, shows the complex fractal structure of a stochastic web. Inside each small island, there is a sub-structure depending upon the value of K_0.

Now let us turn to a more detailed analysis of the mapping $\hat{M}_4$. First, we consider the small values of $K_0 \ll 1$. Let us construct the mapping $\hat{M}_4^4$ in which only the terms containing the first power of K_0 are retained. Quadruple iteration of (6.2.1) yields

$$\hat{M}_4^4 : \begin{cases} \bar{u} = u + 2K_0 \sin \bar{v} \\ \bar{v} = v - 2K_0 \sin u. \end{cases} \tag{6.2.3}$$

The mapping (6.2.3) is written in the form in which it conserves the phase volume. The time interval between two successive steps of the mapping $\hat{M}_4^4$ is 4.

We can easily write the Hamiltonian leading to the mapping (6.2.3)

$$H_4 = -\tfrac{1}{2}K_0 \left\{ \cos v + \cos u \sum_{n=-\infty}^{\infty} \delta(\tfrac{1}{4}\tau - n) \right\} \tag{6.2.4}$$

and the corresponding Hamiltonian equations of motion

$$\dot{u} = \frac{\partial H_4}{\partial v} = \tfrac{1}{2}K_0 \sin v$$

$$\dot{v} = -\frac{\partial H_4}{\partial u} = -\tfrac{1}{2}K_0 \sin u \sum_{n=-\infty}^{\infty} \delta(\tfrac{1}{4}\tau - n). \tag{6.2.5}$$

Fig. 6.2.3 Detailed properties of the phase portrait of the mapping $\hat{M}_4$: (a) a part of the web at $K_0 = 2$; the size of the square $= 4\pi \times 4\pi$; (b) an enlargement of the region inside the small square in (a); the size of the square $= (4\pi \times 4\pi) \times 0.563 \times 10^{-2}$.

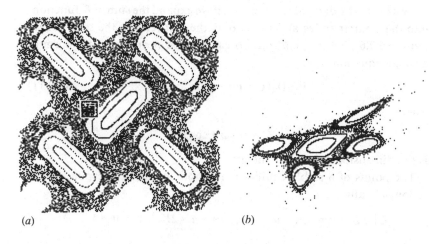

(a)   (b)

Let us single out in (6.2.4) a term with $n = 0$ and rewrite the Hamiltonian H_4 in the following form

$$H_4 = -\Omega_4(\cos v + \cos u) - \Omega_4 \cos u \sum_{\substack{n=-\infty \\ n \neq 0}}^{\infty} \cos(\tfrac{1}{2}\pi n\tau). \qquad (6.2.6)$$

A similar situation was discussed in Sect. 3.4. The first term in H_4 describes the nonlinear oscillations with the following Hamiltonian

$$H_4^{(0)} = -\Omega_4(\cos v + \cos u)$$

$$= -2\Omega_4 \cos\left(\frac{u+v}{2}\right) \cos\left(\frac{u-v}{2}\right) \qquad (6.2.7)$$

and the frequency $\Omega_4 = K_0/2$. The second term in H_4 may be looked at as a perturbation

$$V_4 = -\Omega_4 \cos u \sum_{\substack{n=-\infty \\ n \neq 0}}^{\infty} \cos(\tfrac{1}{2}\pi n\tau). \qquad (6.2.8)$$

Although V_4 has the same order of magnitude as $H_4^{(0)}$ ($\sim K_0$), the minimal frequency of harmonics constituting V_4 is 4. Since $\Omega_4 \ll 4$, the perturbation is a high-frequency one and, as will be shown later, it has a small impact (just as in Sect. 3.4). What is more, similarly to Sect. 3.4, it is sufficient to retain only harmonics with $n = \pm 1$ in (6.2.8), i.e., we assume

$$V_4 \approx -2\Omega_4 \cos u \cos(\tfrac{1}{2}\pi\tau). \qquad (6.2.9)$$

The motion described by the Hamiltonian $H_4^{(0)}$ has already been discussed in Sect. 5.3. From (6.2.7) follow the equations of motion

$$\dot{u} = \Omega_4 \sin v; \qquad \dot{v} = -\Omega_4 \sin u. \qquad (6.2.10)$$

They can also be derived from (6.2.5), if we expand the sum of δ-functions into the Fourier series and retain only the first term in the series as we did in (6.2.6). The system (6.2.10) can also be rewritten in the form of a single equation

$$\ddot{u} + \Omega_4^2(C \sin u - \tfrac{1}{2}\sin 2u) = 0 \qquad (6.2.11)$$

where

$$C = \cos v + \cos u$$

is the dimensionless energy integral.

The points of a stable equilibrium (elliptic points) correspond to the following values

$$|C| = 2; \quad v = \pi n; \quad u = \pi m; \quad m + n = 2l \qquad (l = 0, \pm 1, \dots)$$

while the points of an unstable equilibrium (hyperbolic points) correspond to the values

$$C = 0; \quad v = \pi n; \quad u = \pi m; \quad m + n = 2l + 1 \quad (l = 0, \pm 1, \ldots).$$

Separatrices passing through hyperbolic points cover the phase plane with a square lattice defined by the following equations

$$v = \pm (u + \pi) + 2\pi l \quad (l = 0, \pm 1, \ldots). \tag{6.2.12}$$

Trajectories inside the separatrix cells can be found by integration of equations (6.2.10) and (6.2.11). For $|C| \leq 2$, this yields

$$\cos v = \tfrac{1}{2} C + (1 - \tfrac{1}{2} C) \operatorname{cd}[(1 + \tfrac{1}{2} C)\Omega_4 \tau; \varkappa]$$
$$\cos u = \tfrac{1}{2} C - (1 - \tfrac{1}{2} C) \operatorname{cd}[(1 + \tfrac{1}{2} C)\Omega_4 \tau; \varkappa] \tag{6.2.13}$$

where

$$\varkappa = (2 - C)/(2 + C),$$

while $\operatorname{cd} = \operatorname{cn}/\operatorname{dn}$ is the ratio of the elliptic Jacobian functions.

The trajectories described by the system (6.2.13) are the closed orbits depicted in Fig. 6.2.1*b*. The period of nonlinear oscillations for them is

$$T(C) = \frac{8}{\Omega_4(1 + C/2)} K(\varkappa) \tag{6.2.14}$$

where $K(\varkappa)$ is the full elliptic integral of the first kind. For $C \to 2$, it follows from (6.2.14) that

$$T(2) = 2\pi/\Omega_4,$$

i.e., we get the period of small oscillations in the vicinity of an elliptic point (the expansion of (6.2.7) for small v and u gives the frequency of small oscillations Ω_4).

By taking into account the perturbation (6.2.9), we induce a weak periodic modulation of the whole pattern of the phase plane. This can be seen in Fig. 6.2.1. However, the effect of a perturbation is most vivid near a separatrix: here the destruction of the separatrix takes place with the subsequent formation of a stochastic web. We can consider this process by the high-frequency perturbation (6.2.9) affecting the main motion (6.2.7). It leads to the formation of an exponentially narrow stochastic layer which has a width of the order of $\exp(-\text{const}/K_0)$. This is also the width of the web. A more accurate estimate will be given in the next section.

The existence of an infinite web in our case has the same consequences as in the case of Arnold diffusion. It has a uniform width in the whole phase space and, consequently, an almost uniform velocity of diffusion.

A particle with its initial coordinate falling within the web, can perform a random walk along the web's network moving arbitrarily far from the initial point.

The resonance condition $\omega_4 = \pi/2$ obviously plays an exceptional role in the formation of a web. If this condition is not strictly satisfied, a single separatrix network does not appear even in the zeroth approximation. An example of this is presented in Fig. 6.2.4: here the difference between the phase $\alpha = 1.6$ and its resonance value $\alpha_4 = \omega_4 = 2\pi/4$ is only ~ 0.03. However, a stochastic web may appear even in this case. Provided the perturbation is greater than a certain critical value, it will result in the coupling of isolated motions on the various elements of the disintegrated web.

Inside a separatrix cell, its own set of invariant curves, islands and separatrices forms. The latter are separated from the main network by invariant curves. Thus, the phase portrait inside a cell in many respects resembles one to which the KAM theory is applicable. The form of the portrait to a large extent depends upon the value of K_0. If K_0 increases, the cells of a web become smaller. At the same time, inside a cell, bifurcations of division and the formation of necklaces consisting of various numbers of smaller islets occur. Let us dwell upon this process in more detail.

Fig. 6.2.4   A web's decay due to mismatch of the resonance: $K_0 = 0.7$, $\alpha = \pi/2 + 0.03$; the size of the square $= 12\pi \times 12\pi$.

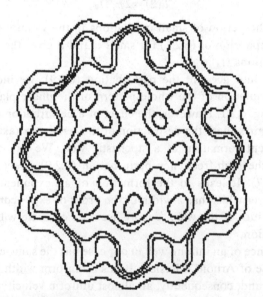

The mapping (6.1.12) always has a stationary point ($u = 0$, $v = 0$). Let us consider matrix $\hat{M}'_\alpha$, tangential to $\hat{M}_\alpha$, in the point $(0, 0)$

$$\hat{M}'_\alpha(0, 0) = \begin{pmatrix} \cos \alpha & K_0 \cos \alpha + \sin \alpha \\ -\sin \alpha & \cos \alpha - K_0 \sin \alpha \end{pmatrix}.$$

Eigenvalues λ of the matrix $\hat{M}'_\alpha(0, 0)$ satisfy an equation

$$\lambda^2 - \lambda \operatorname{Sp} \hat{M}'_\alpha + 1 = 0.$$

Point $(0, 0)$ becomes unstable at $|\operatorname{Sp} \hat{M}'_\alpha| > 2$, i.e., at

$$K_0 > 2 \cot(\alpha/2). \tag{6.2.15}$$

Specifically, in the case of $\alpha = \alpha_4 = \pi/2$, the instability condition has the form of $K_0 > 2$. Instability manifests itself in the fact of point $(0, 0)$ transferring from an elliptic point into a hyperbolic one. This process is accompanied by the birth of two new elliptic points. This is a usual period-doubling bifurcation of the islands (Fig. 6.2.5). Inside an island, we see a new stochastic layer which forms in place of a separatrix passing through saddle $(0, 0)$.

Consider the mapping $\hat{M}_4^2$:

$$\hat{M}_4^2: \begin{cases} \bar{\bar{u}} = -u - K_0 \sin v \\ \bar{\bar{v}} = -v - K_0 \sin(u + K_0 \sin v). \end{cases} \tag{6.2.16}$$

Its stationary points correspond to a cycle with the period 2 and satisfy the following set of equations

$$u_0 = -\tfrac{1}{2} K_0 \sin v_0; \qquad v_0 = -\tfrac{1}{2} K_0 \sin u_0$$

Fig. 6.2.5 Period-doubling bifurcation for $q = 4$ at $K_0 = 2.05$: (a) the size of the square = $2\pi \times 2\pi$; (b) an enlargement of the region inside the square in (a).

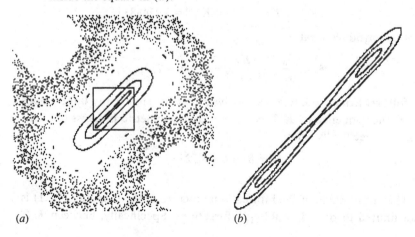

(a)   (b)

or

$$u_0 = \tfrac{1}{2}K_0 \sin(\tfrac{1}{2}K_0 \sin u_0)$$
$$v_0 = \tfrac{1}{2}K_0 \sin(\tfrac{1}{2}K_0 \sin v_0). \qquad (6.2.17)$$

Stability analysis near these points leads us to the following condition

$$|\mathrm{Sp}(\hat{M}_4^2)'| < 2$$

which corresponds to inequality

$$0 < \tfrac{1}{4}K_0^2 \cos(\tfrac{1}{2}K_0 \sin v_0) \cos v_0 < 1. \qquad (6.2.18)$$

Here K_0 and v_0 are related through the condition (6.2.17). It follows from (6.2.18) that a cycle with the period 2 loses stability at the value satisfying an equation

$$\tfrac{1}{2}K_0 \sin \tfrac{1}{2}K_0 = \tfrac{1}{2}\pi,$$

i.e., at $K_0 = 4.88665 \cdots$. This bifurcation corresponds to the formation of a cycle with the period 4.

Let us also note that under the following condition

$$\tfrac{1}{4}K_0^2 \cos(\tfrac{1}{2}K_0 \sin v_0) \cos v_0 = 1$$

an intermediate bifurcation occurs in the system in the case of $K_0 \approx 4.54$. It manifests itself in an appearance of four elliptic points with the former period 2.

As shown in work [1], the loss of stability in a cycle with the period 4 occurs at $K_0 = 4.92934 \cdots$, after which cycles with the period 8 appear. We denote by $K_0^{(n)}$ the value of K_0 corresponding to the period-doubling bifurcations, when the cycle with the period 2^n loses its stability and a cycle with the period 2^{n+1} appears. The sequence of bifurcational values $K_0^{(n)}$ is quickly converging to a limit $K_0^{(\infty)}$

$$2 < K_0^{(1)} < K_0^{(2)} < \cdots < K_0^{(\infty)} = 4.93488 \cdots.$$

Let us introduce a ratio

$$\delta_n = \frac{K_0^{(n-2)} - K_0^{(n-1)}}{K_0^{(n-1)} - K_0^{(n)}}, \qquad n = 2, 3, \ldots.$$

It follows from the numerical analysis that for sufficiently large values of n, the sequence of $K_0^{(n)}$ converges as a geometric progression. This means, specifically, that

$$\lim \delta_n = \delta \approx 8.72 \cdots$$

(Note 6.2).

The entire range of the bifurcational picture of the mapping (6.2.1) is not limited to period-doubling bifurcations. Specifically, between $K_0^{(n)}$

and $K_0^{(n+1)}$, in the process of period-doubling bifurcations, the necklaces consisting of islands corresponding to high-frequency resonances are formed and passed off.

Figure 6.2.6 shows examples of the birth of a group of islands (necklaces) of various types. In other cells of the web, bifurcations occur in much the same manner as in the central one. A complex series of bifurcations forms a 'devil's staircase', which is, as yet, poorly understood.

To end this section, let us note that the cases of $q = 3$ and $q = 6$ also produce a periodic web (Fig. 6.2.7), resembling the classical Koch's fractal (Note 6.3). The lattice formed in this case, is called a 'kagome' lattice. Its properties will be described in the following chapters.

Fig. 6.2.6 Examples of different bifurcations at $q = 4$: (a) the separation of six islands, $K_0 = 2.61$; (b) the separation of thirteen islands, $K_0 = 3.1$; (c) the separation of three islands, $K_0 = 3.7$.

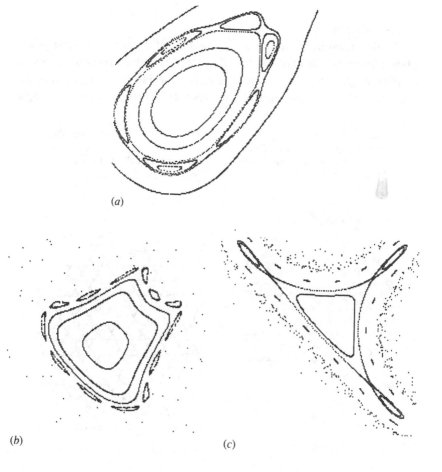

(a)

(b)   (c)

6.3 The aperiodic web and symmetry of plane tilings

We have considered above resonances with $q = 4$ and $q = 3, 6$. They correspond to a phase portrait which, on the phase plane, shows a periodic pattern with a square or hexagonal lattice. The cases of $q = 1, 2$ also produce a periodic – along one axis – phase portrait. Therefore, we can state that a set of values

$$\{q_c\} = \{1, 2, 3, 4, 6\} \tag{6.3.1}$$

is characterized by periodic phase portraits. What happens in the case of other values of q? Before we give the answer to this question, it is useful to look at the initial problem (6.1.15) with the following Hamiltonian

$$H = \tfrac{1}{2}\alpha(u^2 + v^2) - K_0 \cos v \sum_{n=-\infty}^{\infty} \delta(\tau - n) \tag{6.3.2}$$

from a different angle.

The first term in H describes the rotation of a particle. On the phase plane it corresponds to the circles. Therefore, the phase portrait possesses rotational symmetry. It is degenerate, i.e., the family of trajectories on the phase plane is symmetrical with respect to rotation about an angle

Fig. 6.2.7 Two examples of a web at $q = 6$: (a) $K_0 = 0.4$; the size of the square $= 32\pi \times 32\pi$; (b) $K_0 = 0.8$; the size of the square $= 32\pi \times 32\pi$. A network formed on the plane is called the kagome lattice.

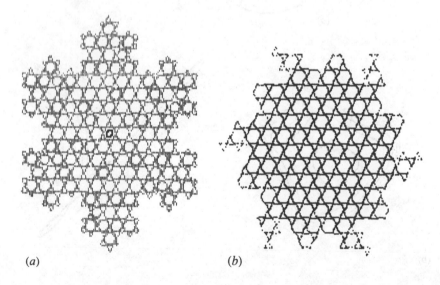

(a)   (b)

$\alpha_q = 2\pi/q$ for an arbitrary integer q. The second term in the Hamiltonian (6.3.2) has translational symmetry with respect to a shift

$$v \rightarrow v + 2\pi m$$

where m is an integer.

Thus, the expression for H contains terms describing two types of motion with rotational and translational symmetry as well as their interaction. For small values of K_0, one could suppose that since perturbation is small, the motion must inherit both types of symmetry. However, there are serious objections against such a simplistic view.

Since the sequence of δ-function-like kicks has the period 1, the perturbation, in the case of $\alpha = \alpha_q = 2\pi/q$, is a resonance one and, therefore, acts most effectively. Hence, one cannot help pondering whether only one symmetry (either rotational or translational) survives in this case, or whether both symmetries continue to coexist. As we have already seen, when $q \in \{q_c\}$, the second case holds true. All obtained patterns of phase portraits possess both translational and rotational symmetry. However, we know that such coexistence of symmetries on the plane, for arbitrary patterns, is possible only in the cases defined by (6.3.1) [4, 5]. This statement can be proved very easily.

Let the points A and B (Fig. 6.3.1) be the nodes of a certain network covering the plane. Also let this cover be periodic and possess symmetry with respect to the translation by the period a. If the points A and B are adjacent, the distance between them is a. If, at the same time, the network has a rotational symmetry over the angle $\alpha = 2\pi/q$, the axes of this symmetry must pass through both of these points A and B. Let us perform a rotation about the angle α around the point A. In so doing, the point B turns into B' (Fig. 6.3.1). Similarly, a rotation about the angle α around the point B, turns the point A into A'. The distance $A'B'$ must be divisible

Fig. 6.3.1 The determination of conditions for the coexistence of translational and rotational symmetries.

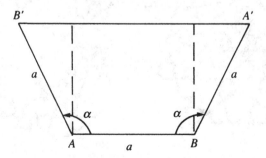

by the period of the lattice a, i.e., $A'B' = ma$, where m is an integer. Otherwise, translational symmetry should be broken. It also follows from Fig. 6.3.1 that

$$A'B' = ma = a + 2a \sin(\alpha - \pi/2)$$

or

$$\cos \alpha = (1 - m)/2.$$

Hence, for m, the following values: $m = 3, 2, 1, 0$ are possible. Respectively, for $q = 2\pi/\alpha$, the values $q = 2, 3, 4, 6$ are possible. If we add the trivial case of $q = 1$, we arrive at the set $\{q_c\}$ in (6.3.1).

Thus, for values of q not belonging to the set $\{q_c\}$, there can be no periodic web. However, there can exist an aperiodic web with rotational symmetry. Such patterns have been discovered in quasi-crystals and will be discussed in more detail later. The examples of webs for $q = 5, 7, 8$ are presented in Figs. 6.3.2 and 6.3.3. In all these examples, the web is formed by random walks of the point on the phase plane. Since the point cannot travel beyond the web's channels, after a sufficiently long time we possess rather detailed information about the web's structure.

The structure of a web develops rather unevenly in time. In Fig. 6.3.2a the web is given as a resut of a substantial time-development (computation time). A quite different structure is displayed in Fig. 6.3.2b. After about 10^6 steps of the mapping, only a window in the form of a fractal

Fig. 6.3.2   A stochastic web at $q = 5$ for two different initial conditions (see text): $K_0 = 0.7$; the size of the square $= 256\pi \times 256\pi$.

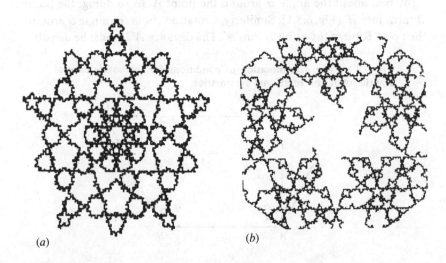

(a)   (b)

pentagonal star is left within the web. The time needed by the web to cover this window is $\tau > 1.3 \times 10^6$. Different shapes and sizes of windows depend on the choice of initial conditions. The dynamics of a 'snowflake' shaping out of the web is very sensitive to initial conditions because of the very existence of the Cantori suppressing diffusion in certain directions.

How does the web grow? First, the random walks of a particle on the plane create a certain irregular figure – for example a star the ends of which have grown to different lengths. Next, the star acquires a regular form, after which the formation of a larger star begins (Fig. 6.3.4). Of course, the boundaries of these stars have a complex shape and form fractal curves as $t \to \infty$. Thereby, a stochastic web can also be called a fractal web.

A fractal web has a central free window in the form of a regular q-square for even q or a regular $2q$-square for uneven q. Inside the central window also lie stochastic layers; however, they do not expand beyond the window's boundary to unite with the main web (Fig. 6.3.5). The smaller the value of K_0, the greater is the size of the central window. The web has an infinite number of elements of same shape as the central window. These elements form an aperiodic tiling of the plane (Fig. 6.3.6).

If $K_0 \to 0$, the width of the web diminishes. We have already shown this for the case of $q = 4$, while for $q \notin \{q_c\}$ the corresponding estimations shall be given later. A periodic web with $q \in \{q_c\}$ shall be called a web

Fig. 6.3.3 A stochastic web in the case of $q = 7$ and $q = 8$: (a) $q = 7$, $K_0 = 0.5$; the size of the square $= 64\pi \times 64\pi$; (b) $q = 8$, $K_0 = 0.6$; the size of the square $= 128\pi \times 128\pi$.

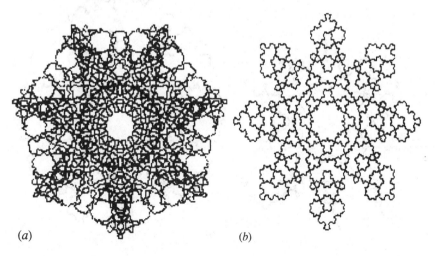

(a)   (b)

with crystal symmetry, while the web with $q \notin \{q_c\}$ shall be called a web with quasi-crystal symmetry, or simply quasi-symmetry. The change in K_0 does not affect the size of the cells of the web with crystal symmetry. In the case of quasi-symmetry this does not hold true. One may expect the web with $q \notin \{q_c\}$ to exist even at $K_0 \to 0$.

Let us mention a special observation concerning the mapping $\hat{M}_q$. It generates a certain invariant set on a plane – a stochastic web. This set has an almost regular shape and realizes a plane tiling with an arbitrary

Fig. 6.3.4 Four consecutive steps of the formation of a web with $q = 5$: $K_0 = 0.75$; the size of the square $= 80\pi \times 80\pi$.

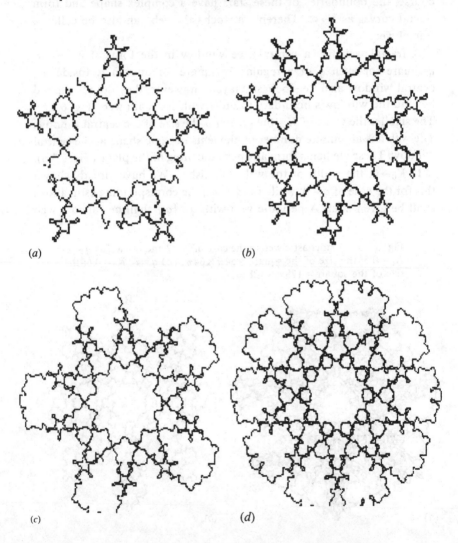

(a)   (b)

(c)   (d)

Fig. 6.3.5   The phase portrait of the interior of the central window of a web with $q = 5$: $K_0 = 0.75$; the size of the square $= 10\pi \times 10\pi$.

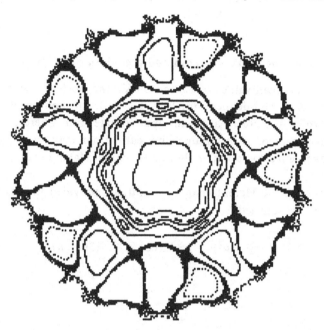

Fig. 6.3.6   A web with $q = 5$ has a repeating element in the form of the central window: $K_0 = 0.7$; the size of the square $= 100\pi \times 100\pi$.

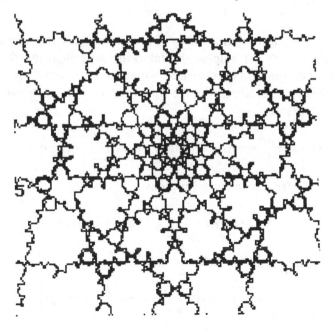

qth-order symmetry. Therefore, the mapping $\hat{M}_q$ may be called a generator of tilings with qth symmetry. The properties of these tilings will be discussed in the next chapter.

6.4 The web's skeleton and the width of the web

In the case of small values of parameter K_0, the stochastic web is very thin and has a well-defined structure. Let us try to find an algorithm defining this structure in a certain approximate form where all minor details are omitted. Such an algorithm can be obtained by means of a certain averaging operation applied to the initial Hamiltonian in (6.3.2) [11, 3].

Let us write the Hamiltonian of a particle, provided the resonance conditions are satisfied

$$H = \tfrac{1}{2}\alpha_q(u^2 + v^2) - K_0 \cos v \sum_{n=-\infty}^{\infty} \delta(\tau - n) \qquad (6.4.1)$$

where

$$\alpha_q = 2\pi/q$$

and q is an integer. Let us express (u, v) and I in polar coordinates

$$u = \rho \cos\theta; \qquad v = -\rho \sin\theta; \qquad I = \tfrac{1}{2}\rho^2$$

where ρ is the dimensionless radius of the particle's rotation (the cyclotron radius). With the help of the following generating function $F = (\theta - \alpha_q\tau)J$ we may transfer to new variables $J = I$, $\varphi = \theta - \alpha_q\tau$, in a coordinate system rotating with the frequency α_q. In the new variables, the Hamiltonian is

$$\tilde{H} = H + \frac{\partial F}{\partial \tau} = -K_0 \cos[\rho \sin(\varphi + \alpha_q\tau)] \sum_{n=-\infty}^{\infty} \delta(\tau - n). \qquad (6.4.2)$$

Let us transfer a series of δ-functions

$$\sum_{n=-\infty}^{\infty} \delta(\tau - n) = \sum_{j=1}^{q} \sum_{m=-\infty}^{\infty} \delta[\tau - (mq + j)]. \qquad (6.4.3)$$

We shall make use of the following notation

$$\sum_{m=-\infty}^{\infty} \delta(\tau - j - mq) = \frac{1}{q} \sum_{m=-\infty}^{\infty} \exp\left\{\frac{2\pi i m}{q}(\tau - j)\right\}.$$

By substituting it into (6.4.3) and (6.4.2) we get

$$\tilde{H} = H_q + V_q$$

$$H_q = -\frac{1}{q} K_0 \sum_{j=1}^{q} \cos(\xi_j) \qquad (6.4.4)$$

$$V_q = -\frac{1}{q} 2K_0 \sum_{j=1}^{q} \cos(\xi_j) \sum_{m=1}^{\infty} \cos\left[\frac{2\pi m}{q}(\tau - j)\right]$$

where

$$\xi_j = -\rho \sin\left(\varphi + \frac{2\pi}{q}j\right) = v \cos\left(\frac{2\pi}{q}j\right) - u \sin\left(\frac{2\pi}{q}j\right). \qquad (6.4.5)$$

According to definition (6.4.5), we can also write

$$\xi_j = \boldsymbol{\rho} \boldsymbol{e}_j$$

$$\boldsymbol{e}_j = \left(\cos\frac{2\pi}{q}j, -\sin\frac{2\pi}{q}j\right); \qquad \boldsymbol{\rho} = (v, u), \qquad (6.4.6)$$

i.e., $\boldsymbol{e}_j$ is a unit vector defining the jth vertex of a regular q-star.

We shall call the expression

$$H_q = -\tfrac{1}{2}\Omega_q \sum_{j=1}^{q} \cos(\boldsymbol{\rho} \boldsymbol{e}_j); \qquad \Omega_q \equiv \frac{2K_0}{q} \qquad (6.4.7)$$

the resonance Hamiltonian of the order of q. For $q = 4$, it describes the already familiar expression

$$H_4 = -\Omega_4(\cos u + \cos v). \qquad (6.4.8)$$

A phase portrait corresponding to the system (6.4.8) is depicted in Fig. 6.4.1. It shows the lines of the level of a surface

$$\mathscr{E}_4(u, v) = -H_4/\Omega_4.$$

For $\mathscr{E}_4 = 0$, they form a square lattice. The values of $\mathscr{E}_4 > 0$ correspond to the humps of the surface $\mathscr{E}_4(u, v)$, while the values $\mathscr{E}_4 < 0$ correspond to its valleys. They are arranged as on a chess board.

Let us see in more detail how perturbation V_4 leads to the formation of a stochastic web [1, 3, 12].

As we have shown above, there is an accurate transition from the initial Hamiltonian in (6.4.1) to the Hamiltonian $\tilde{H}$ which, in the case of $q = 4$ assumes the form of

$$\tilde{H} = H_4 + V_4.$$

For V_4, we have

$$V_4 = -2\Omega_4(\cos v - \cos u)\cos \pi\tau + \cdots \qquad (6.4.9)$$

where the ellipses denote all terms in the sum over m in (6.4.4), beginning with $m = 3$. We shall neglect them since they make too small a contribution to the width of the stochastic layer which forms due to the first term in V_4.

Equations (6.4.8) and (6.4.9) give us

$$\dot{H}_4 = 4\Omega_4^2 \sin u \sin v \cos \pi\tau. \qquad (6.4.10)$$

The equation of separatrices found from (6.4.8) for $H_4 = 0$, has the following form

$$v = \pm(u + \pi) + 2\pi m \qquad (m = 0, \pm 1, \ldots).$$

The solution on separatrices has the following form

$$\sin u = -\sin v = -\frac{1}{\cosh \Omega_4(\tau - \tau_n)} \qquad (6.4.11)$$

Fig. 6.4.1 The phase portrait on the plane (u, v) for the resonance Hamiltonian of the fourth order. The closed orbits correspond to the values $\mathscr{E}_4 = 1, -1, -3$.

where τ_n is a constant defining the point of reference in time. Substituting (6.4.11) into (6.4.10) and performing the integration, we find the energy change under the effect of a perturbation in the vicinity of a separatrix

$$\Delta H_4 = -4\Omega_4^2 \int_{-\infty}^{\infty} d\tau \frac{\cos \pi\tau}{\cosh^2 \Omega_4'(\tau - \tau_n)} = -4\pi^2 \frac{\cos \pi\tau_n}{\sinh(\pi^2/2\Omega_4)}.$$

Since $\Omega_4 = K_0/2 \ll 1$, this expression assumes the following form

$$\Delta H_4 = -8\pi^2 \cos \pi\tau_n \exp(-\pi^2/K_0). \tag{6.4.12}$$

The period of oscillations in the vicinity of a separatrix is (see also (6.2.14))

$$T(H_4) = \frac{4}{\Omega_4} \ln(8\Omega_4/|H_4|) \tag{6.4.13}$$

if we set $\varkappa = 1 - C = 1 - (H_4/\Omega_4)$. The time needed to pass near a separatrix is equal to one quarter of the period, i.e.,

$$\tau_{n+1} - \tau_n \approx T(H_4)/4 = \frac{1}{\Omega_4} \ln(8\Omega_4/|H_0|).$$

Putting H_4 in this formula equal to H_{n+1} and taking the expression (6.4.12) into account, we get the mapping near a separatrix

$$H_{n+1} = H_n - 8\pi^2 \exp(-\pi^2/K_0) \cos \pi\tau_n$$

$$\tau_{n+1} = \tau_n + \frac{1}{\Omega_4} \ln(8\Omega_4/|H_{n+1}|). \tag{6.4.14}$$

This procedure has already been described in Sect. 3.1 and 3.4. Now following the same procedure, we find the border of the stochastic layer. It can be obtained from the following condition

$$\max \left| \frac{\delta\tau_{n+1}}{\delta\tau_n} - 1 \right| > 1.$$

Hence, for the border H_s, we have

$$H_s = \frac{16\pi^3}{K_0} \exp(-\pi^2/K_0). \tag{6.4.15}$$

Quantity $2H_s$ defines the width of a layer. Since all separatrices merge forming a single network - a web - quantity $2H_s$, at the same time, is the width of the stochastic web.

Now consider the case of $q = 3$ or 6. It follows from (6.4.4) that

$$H_3 = H_6 = -\tfrac{1}{2}\Omega_3 \left[\cos v + \cos \left(\frac{1}{2}v + \frac{\sqrt{3}}{2}u \right) + \cos \left(\frac{1}{2}v - \frac{\sqrt{3}}{2}u \right) \right] \tag{6.4.16}$$

$$\Omega_3 = 2K_0/3.$$

The phase portrait for the Hamiltonian $\mathscr{E}_3 = -2H_3/\Omega_3$ is shown in Fig. 6.4.2. A separatrix network for $\mathscr{E}_3 = -1$, forms the kagome lattice. It is defined by equations

$$v = \pi(2n_1+1) \qquad (n_1 = 0, \pm 1, \ldots)$$

$$v = \sqrt{3}\,u + 2\pi(2n_2+1) \qquad (n_2 = 0, \pm 1, \ldots) \qquad (6.4.17)$$

$$v = -\sqrt{3}\,u + 2\pi(2n_3+1) \quad (n_3 = 0, \pm 1, \ldots).$$

They correspond to three independent families of parallel and equally spaced straight lines.

On all separatrices (6.4.17), the energy integral H_3 assumes the same meaning $\mathscr{E}_3 = -1$.

Within the range of values $-\frac{3}{2} \leqslant \mathscr{E} < -1$, a particle moves inside the triangles of the lattice. For $-1 < \mathscr{E}_3 \leqslant 3$, motion occurs within hexagons.

The width of a stochastic web can be obtained in the same way, since perturbation V_3 has the same structure as V_4:

$$H_s \sim \exp(-2\sqrt{3}\,\pi^2/3K_0).$$

Fig. 6.4.2 The phase portrait on the plane (u, v) for the resonance Hamiltonian of the third order. The closed orbits correspond to the values $\mathscr{E}_3 = 2$ (inside the hexagons) and $\mathscr{E}_3 = -\frac{3}{2}$ (inside the triangles).

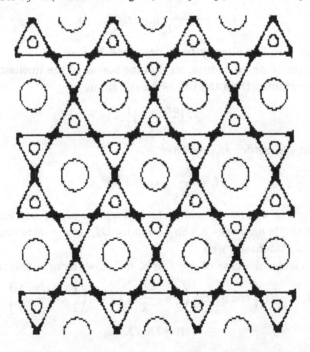

Similar considerations can be used in the case of an arbitrary value of q, so that we can write the width of a stochastic web in the following form

$$H_s \sim \exp(-\text{const}/K_0) \qquad (6.4.18)$$

where the value of const increases somewhat with the growth of q. However, for all $q \notin \{q_c\}$, there is a singularity which we must discuss in more detail.

There are a number of basic differences between the properties of the Hamiltonians H_q in the case of $q \in \{q_c\}$ and in the case of $q \notin \{q_c\}$. If $q \in \{q_c\}$, all separatrices of a dynamic system with the Hamiltonian H_q belong to the same value of the energy integral $\mathcal{E}_q = \mathcal{E}_q^{(0)}$, where

$$\mathcal{E}_q = -2H_q/\Omega_q = \sum_{j=1}^{q} \cos(\boldsymbol{\rho}\boldsymbol{e}_j). \qquad (6.4.19)$$

As we have already seen above

$$\mathcal{E}_4^{(0)} = 0; \qquad \mathcal{E}_3^{(0)} = -1.$$

This circumstance also means that all hyperbolic points are situated on the same plane of the energy level $\mathcal{E}_q^{(0)}$, for $q \in \{q_c\}$. In the case of $q \notin \{q_c\}$, this property is absent. Saddles and separatrices are distributed among different planes of the level, this being one of the structural differences between periodic patterns and patterns with quasi-symmetry. In the case of $q \notin \{q_c\}$, there is no single network of separatrices. The plane depicting the function defined by equation (6.4.19) is presented in Fig. 6.4.3 for $q = 5$. This figure shows that saddles form a pattern resembling a family of parallel straight lines. However, in Fig. 6.4.4 we see that, near the value of $\mathcal{E}_5 = 1$, separatrix loops come very close to each other. Still, the loops intersect in various points, situated on different (although close) lines of the energy level.

Due to fact that on the surface of the energy integral $\mathcal{E}_5 \approx 1$ the gaps between some separatrices are very small, even a small perturbation can create stochastic layers which cover these gaps. Thus, a single stochastic web is formed. As we see, in the case of quasi-symmetry, the mechanism of the web's formation sufficiently differs from that of crystal symmetry, since if we neglect the perturbation V_q, there is no single separatrix network.

If we consider the phase volume presented as a thin layer of the width of $\Delta\mathcal{E}$ in the vicinity of the value $\mathcal{E}_q = \mathcal{E}_q^{(0)}$, all lines of the level (including separatrices), satisfying the following equation

$$\mathcal{E}_q(u, v) = \text{const} = \mathcal{E}_q^{(0)}$$

will be smeared and assume a finite thickness. Let us choose the value of $\mathscr{E}_q^{(0)}$ so that all separatrix loops come sufficiently close to each other. For example, for $q = 5$, this happens at $\mathscr{E}_q^{(0)} \approx 1$. Then, because of a finite width of separatrices, they merge forming a certain pattern (network) on the phase plane. Fig. 6.4.4c shows such a pattern. Let us call these patterns the web's skeleton. The skeleton is a certain approximation of a stochastic web for the initial Hamiltonian $H(u, v, \tau)$. The analysis of the web's skeletons will be made in the next chapter.

6.5 Patterns in the case of a particle's diffusion

So far, we have discussed only small values of the parameter of perturbation K_0. Only for this case, the structure of a thin web generated by the mapping $\hat{M}_q$ did not change. As parameter K_0 grows, however, the web becomes very wide. Small details are destroyed. The further increase of

Fig. 6.4.3   The surface $\mathscr{E}_5(u, v)$ in the case of quasi-symmetry of the fifth order.

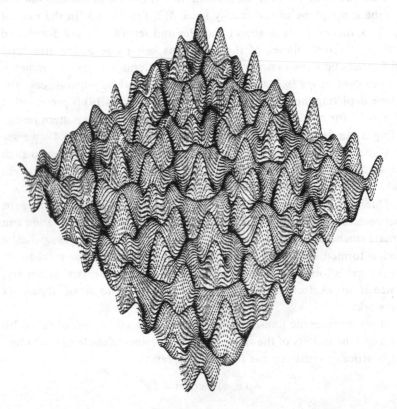

Fig. 6.4.4   The phase portrait for the Hamiltonian $\mathscr{E}_5$ with fifth-order quasi-symmetry: (*a*) the lines of the level near $\mathscr{E}_5 = 1$; all points belong to the range of values $0.8 < \mathscr{E}_5 < 1.2$; the size of the square = $32\pi \times 32\pi$; (*b*) an enlarged image of lines of the level in the small square marked in (*a*) $0.9 < \mathscr{E} < 1.1$; (*c*) the lines of the level in the vicinity of $\mathscr{E}_5 = 0.2$ (thick lines) and in the vicinity of $\mathscr{E}_5 = -3.2$ (thin lines); the size of the square = $32\pi \times 32\pi$.

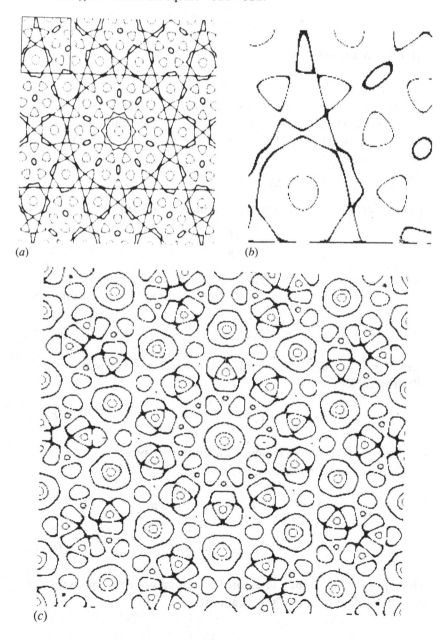

(*a*)

(*b*)

(*c*)

K_0 produces a pattern formed by the random walks of a particle on the plane which has nothing in common with the pattern of a periodic and aperiodic web in the case of small K_0.

Let us turn to the initial Hamiltonian in the problem of a particle's motion in a magnetic field and the field of a wave packet (6.1.14)

$$H = \tfrac{1}{2}(\dot{x}^2 + \omega_0^2 x^2) - \frac{e}{m} E_0 T \cos kx \sum_{n=-\infty}^{\infty} \delta(t - nT). \qquad (6.5.1)$$

Let us introduce the action-angle variables

$$I = \dot{x}^2 + \omega_0^2 x^2; \qquad \varphi = \tan^{-1}(\omega_0 x / \dot{x}).$$

The quantity I, with an accuracy up to a constant factor, is equal to the energy of a particle (oscillator). In the case of sufficiently large values of I, the relative change of the action due to one kick of a δ-function will be small, i.e.,

$$|\Delta I| = |I_{n+1} - I_n| \ll I_n. \qquad (6.5.2)$$

If at the same time, the following condition is satisfied

$$K_0 = \frac{e E_0 kT}{m \omega_0} \gg 1 \qquad (6.5.3)$$

(see the definitions (6.1.8) and (6.1.13)), the motion of a particle is highly stochastic and can be described by the equation of Fokker–Plank diffusion

$$\frac{\partial F}{\partial t} = \frac{1}{2} \frac{\partial}{\partial I} D(I) \frac{\partial F}{\partial I}. \qquad (6.5.4)$$

Here $F = F(I, t)$ is the distribution function and $D(I)$ is a particle's diffusion coefficient

$$D(I) = \frac{1}{T} \langle\!\langle (\Delta I)^2 \rangle\!\rangle \qquad (6.5.5)$$

while the brackets $\langle\!\langle \cdots \rangle\!\rangle$ denote an average over phase φ.

It follows from the mapping $\hat{M}_\alpha$ (6.1.12)

$$\Delta I = \frac{2e}{m} E_0 T I^{1/2} \sin \varphi \sin(k\rho \cos \varphi)$$

$$+ \left(\frac{e}{m \omega_0} E_0 kT \right)^2 \sin^2(k\rho \cos \varphi) \qquad (6.5.6)$$

where $\rho = I^{1/2}/\omega_0$ is the radius of the rotation orbit of a particle (the amplitude of an oscillator's oscillations).

The substitution of the expression (6.5.6) into (6.5.5) and integration over phase φ yield

$$D(I) = \frac{\omega_0^2 K_0^2 I}{k^2 T}\left[1 - \frac{1}{k\rho} J_1(2k\rho)\right] \qquad (6.5.7)$$

where J_1 is the Bessel function. In the process of averaging over phase, the information on symmetry of the problem is lost. In practice, it is still available if the values of q are close to rational numbers. In Figs. 6.5.1–6.5.3 we see examples of how a particle's trajectory fills the phase plane for $q = 3$, $q = 5$, $q = 7$. Figures formed in this way are typical fractals. Their shape is highly dependent upon the time-development of diffusion and upon parameter K_0. The larger K_0, the more the fractals are elongated along the radii. A fractal tree appears on the phase plane. Various forces presented in the equations of motion lead to a certain clustering of diffusion. In the case of large amplitudes of the perturbation ($K_0 \gg 1$), a particle rapidly moves along a radius, creating smeared rays of an irregular shape. The smearing of rays in the azimuthal direction is a relatively slow process. Therefore, if we choose the initial condition so that diffusion is directed mostly along other radii, the new fractal tree (claster) during a considerable time will not intersect with the former tree. Thus, the phase plane is 'clasterized', i.e., it is partitioned into certain fractal domains of diffusion which only slightly overlap.

Fig. 6.5.1 The fractal generated by diffusion with third-order symmetry at $K_0 = 3.5$: (a) the size of the square $= 100\pi \times 100\pi$; the number of steps $= 1.5 \times 10^5$; (b) the sector contained in the small square in (a); the number of steps $= 9500$.

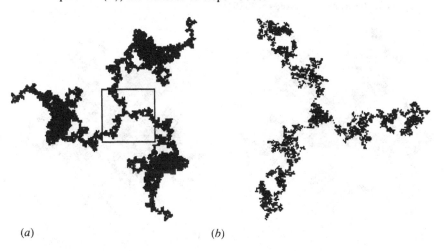

(a)   (b)

6.6 The breaking up of the web in the case of relativistic particles

Suppose that a particle travels in a magnetic field and in the field of a wave packet, perpendicular to the magnetic field (see the initial problem (6.1.5)). Then, provided the resonance condition

$$\omega_0 T = \alpha = 2\pi/q, \qquad (q = 3, 4, \ldots)$$

and $\omega = 0$, a stochastic web appears. Generally speaking, an infinite acceleration of particles is possible along the channels of the web. However, when the values of energy are very high, the particle's velocity approaches the velocity of light. The particle's motion becomes relativistic, while the frequency of its rotation in the magnetic field becomes nonlinear, i.e., depends upon the particle's energy. As a result, the web must break up in the region of a sufficiently high nonlinearity. This effect was studied in work [13] (Note 6.4, see also Note 6.5).

In a relativistic case, equation (6.1.5) transfers to the following equation

$$\frac{d}{dt}(\gamma\dot{x}) + \frac{\omega_0^2}{\gamma}x = \frac{e}{m}E(x, t)$$

$$\dot{y} = -\frac{\omega_0}{\gamma}x,$$

(6.6.1)

where γ is the relativistic factor

$$\gamma = [1 - (\dot{x}^2 + \dot{y}^2)/c^2]^{1/2}, \qquad (6.6.2)$$

while the wave packet is defined by the same expression (6.1.3) where

Fig. 6.5.2   The fractal generated by diffusion with fifth-order symmetry at $K_0 = 3.5$: (*a*) the size of the square $= 800\pi \times 800\pi$; the number of steps $= 160\,000$; (*b*) the sector inside the small square in (*a*); the number of steps $= 9826$.

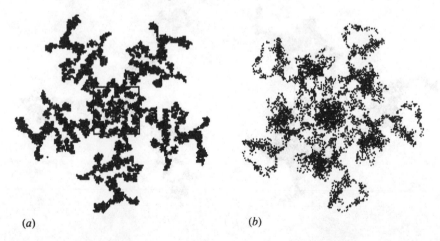

(*a*)   (*b*)

$\omega = 0$, and

$$E(x, t) = -E_0 T \sin kx \sum_{n=-\infty}^{\infty} \delta(t - nT). \qquad (6.6.3)$$

The substitution of (6.6.3) and (6.6.2) into (6.6.1) enables us to transfer from the differential equation to the mapping. With this aim in view, we integrate (6.6.1) in the vicinity of the δ-function then match the variables $(\dot{x}_n, \dot{y}_n)$ and $(\dot{x}_{n+1}, \dot{y}_{n+1})$ at the ends of the interval of time of the length T, in the same way as in Sect. 6.1. Finally, we arrive at the

Fig. 6.5.3 The fractal generated by diffusion with seventh-order symmetry at $K_0 = 3.5$: (a) the size of the square $= 800\pi \times 800\pi$; the number of steps $= 108\,470$; (b) a sector inside the small square in (a); the number of steps $= 7618$.

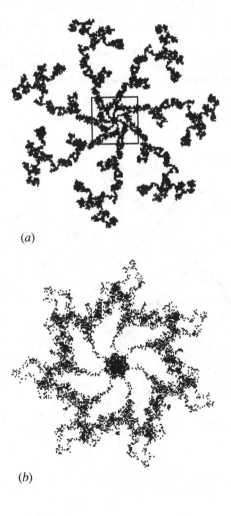

(a)

(b)

relativistic mapping obtained in [13]

$$\hat{M}_\alpha(\gamma): \begin{cases} u_{n+1} = (u_n + K_0 \sin v_n) \cos(\alpha/\gamma_n) + v_n \sin(\alpha/\gamma_n) \\ v_{n+1} = -(u_n + K_0 \sin v_n) \sin(\alpha/\gamma_n) + v_n \cos(\alpha/\gamma_n) \end{cases} \quad (6.6.4)$$

where we denote

$$\gamma_n = \left\{ 1 + \frac{\omega_0^2}{k^2 c^2} [(u_n + K_0 \sin v_n)^2 + v_n^2] \right\}^{1/2}$$

$$u = \gamma k\dot{x}/\omega_0; \qquad v = \gamma k\dot{y}/\omega_0 = -kx \quad (6.6.5)$$

$$K_0 = \frac{eE_0 kT}{m\omega_0}.$$

At $\gamma = 1$, the mapping $\hat{M}_\alpha(\gamma)$ is reduced to (6.1.12).

Equations (6.6.4) and (6.6.5) show differences caused by the relativism of particles. In place of the twisting angle α, now in $\hat{M}_\alpha(\gamma)$ we see the

Fig. 6.6.1   The separatrices of the averaged Hamiltonian $H_4^{(0)}(\gamma)$ in the case of a fourth-order resonance and $\Gamma = 1.3 \times 10^{-4}$. Dotted lines correspond to a separatrix network at $\Gamma = 0$ (Fig. in Longcope and Sudan [13]).

angle α/γ_n. Consequently, the resonance condition now has the following form

$$\alpha/\gamma_n = 2\pi/q \qquad (q = 1, 2, \ldots). \qquad (6.6.6)$$

The value of γ_n depends upon the particle's energy. For that reason, condition (6.6.6) can be satisfied in the case of certain values of (u, v). Variations of (u, v) in the course of time lead to a detuning of the resonance. Degeneracy is thereby eliminated, so that we are able to apply the results of the KAM theory concerning the conservation of invariant curves embracing the centre if the perturbation is small.

To illustrate this conclusion, let us consider the case of a fourth-order resonance ($q = 4$). By expanding the expressions in (6.6.4)–(6.6.6) in small values of parameters $\beta = \omega_0/kc$ and K_0, we get the following mapping

$$\hat{M}_4^4(\gamma): \begin{cases} u_{n+4} = u_n + 2K_0 \sin v_n - \pi\beta^2(u_n^2 + v_n^2)v_n \\ v_{n+4} = v_n - 2K_0 \sin u_n + \pi\beta^2(u_n^2 + v_n^2)u_n, \end{cases} \qquad (6.6.7)$$

Fig. 6.6.2 Five different initial conditions for the relativistic mapping $\hat{M}_4^4(\gamma)$: $\beta = 0.01$; $K_0 = 1.2$ (Longcope and Sudan [13]).

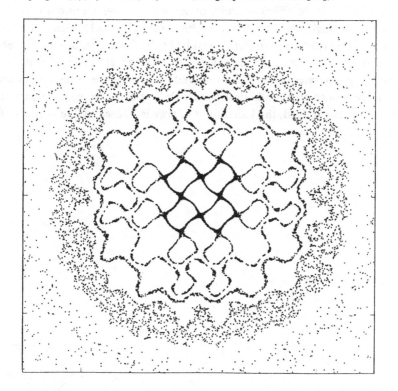

with an accuracy up to first-order terms in K_0 and β^2 [13]. If we compare this mapping to the mapping $\hat{M}_4^4$ (6.2.3) we see that the latter does not contain the last terms with β^2. Further processing is analogous to that performed in Sect. 6.2. The Hamiltonian leading to (6.6.7) has the following form

$$H_4(\gamma) = -\tfrac{1}{2}K_0\left\{[\cos v + \cos u + \Gamma(u^2+v^2)^2]\right.$$

$$\left. + [\cos u + \Gamma(u^2+v^2)^2] \sum_{\substack{n=-\infty \\ n \neq 0}}^{\infty} \cos\left(\frac{\pi n t}{2T}\right)\right\} \qquad (6.6.8)$$

where

$$\Gamma = \pi\beta^2/8K_0.$$

Expression (6.6.8), in turn, should be compared to the corresponding formula (6.2.6), which follows from (6.6.8) in the case of $\Gamma = 0$ ($\beta = 0$). The unperturbed part of the Hamiltonian $H_4(\gamma)$ has the following form

$$H_4^{(0)}(\gamma) = -\Omega_4[\cos v + \cos u + \Gamma(u^2+v^2)^2] \qquad (6.6.9)$$

(see (6.2.7)). It describes a family of separatrices, which in our case do not form a single network, covering the plane (Fig. 6.6.1). Therefore, the effect of a perturbation in (6.6.8) results in the formation of a finite web (Fig. 6.6.2). This example also shows that the acceleration of particles up to high energies is possible, provided we choose the initial condition in the region of an already destroyed web. In the region where a part of the web has survived, the increase of energy is limited by the borders of the remaining web.

PART III · Spatial patterns

7
Two-dimensional patterns with quasi-symmetry

The physical phenomena taking place in condensed matter are far more diverse than phenomena produced by the motion of discrete particles. As the number of degrees of freedom increases, the need arises for a qualitatively new analysis of the dynamics of matter. One of the important manifestations of the difference between these two cases is the formation of regular patterns in matter. Their examples are many and varied: crystals and vortex streets, convective cells and high clouds, fracture patterns and soap foam, the arrangement of seeds in a sunflower head and Jupiter's Great Red Spot, snowflake patterns and the 'packing' of live cells. The list could be continued indefinitely. It is only natural that we should yearn for universality in understanding all these phenomena. Following the recent discovery of quasi-crystals, we understand much more about the possible types of patterns and symmetries. Speaking of the symmetry of some particular object, we imply that it possesses a certain invariant property. Although in the analysis of spatial patterns and their symmetries, the use of geometrical methods is quite natural, dynamic methods might come in useful as well.

As demonstrated above, the dynamics of particles and fields can also exhibit certain symmetries. In this chapter we shall learn in which way symmetry in dynamic problems can help achieve and analyse symmetry in spatial patterns. For various reasons, the existence of this relation proves very fruitful and enables deeper penetration into the nature of things. Recall what efforts and ingenuity it took Kepler to convince his readers of the existence of certain formative forces and factors. Spatial patterns and laws of their growth are determined by interaction (Note 7.1).

Penetrating deep into the dynamic processes preceding the emergence of patterns, we are now able to apply new methods of nonlinear dynamics to the analysis of geometrical properties of patterns where before it seemed impossible. The most interesting subtlety of this technique lies in the possibility of introducing a basically new element – namely, weak chaos in the geometry of patterns – into the method of analysis. In the general case, chaos is an irremovable property of dynamics. Therefore, forms of dynamic origin can also include small stochastic deviations from the ideal case. Thus, dynamic stochasticity might give a clue to the way small geometrical discrepancies can be compensated for in patterns which in principle do not exist in the ideal form.

7.1 What types of patterns are there?

When we try to understand why, for example, snowflakes have a hexagonal shape instead of, for example, a pentagonal one, we feel the need to penetrate into the origins of the geometrical features of natural objects. Do geometrical laws rule out the existence of certain forms of physical objects? In one way or another, these questions were first asked in ancient times. Plato's regular bodies played an important role in Kepler's picture of the Universe. Today the laws of symmetry in physics have become a common method of analysis.

The use of geometrical ideas is most graphic in crystal physics. The key to orthodox crystallography is the idea of periodically recurrent patterns filling in a space or a plane. In a more formal sense, in crystals we see the packing of one or several structural cells which have translational symmetry with respect to the shift by a certain vector. In such cases, we speak of long range order in crystals.

Symmetry properties of objects possess universality. This means that, if a method of regular packing of cells of a certain shape has been found in crystals, a similar way of packing 'liquid' cells can be discovered in hydrodynamic flows and in the structure of the phase plane of a dynamic system. Therefore, the problem of mosaics or tilings in space or on a plane, possessing certain symmetry, is connected not only to the geometrical properties of space but also to real physical processes (Note 7.2).

Although it is possible to give a very complex shape to a single cell of a mosaic pattern (as proved by the ingenious drawings of the Dutch painter Maurits Escher [9]), still there is a certain limited number of ways to cover our plane with these cells if we want to retain the long range order. The importance of this problem – that of paving an infinite plane with only a finite number of elements – was well understood by Kepler. This is proved by his studies reflected in Fig. 7.1.1.

In all the research on tilings, packings, mosaics and ornaments, the number 5 always stood in a class by itself. Attempts to understand whether a pentagon might be used in pavements can be found in Kepler's (see Fig. 7.1.1, element Aa) and Dürer's works. Although Muslim ornaments with regular pentagons and decagons (see Sect. 10.2) have been known for a long time, in specialist literature it was typical to claim that crystals with a fifth-order symmetry axis cannot exist.

However, attempts were made in crystallography to break with orthodox views on the structure of crystals. The first convincing argument in favour of this was given by Schrödinger [10]. In order to explain the existence of the regular structure of the giant molecule constituting a gene, he introduced the notion of an aperiodic crystal. The genetic code is the algorithm determining the sequence of atoms and atomic groups in aperiodic crystals. Many attempts to extend the old notions of order in crystals were centred on the search for patterns with fifth-order symmetry. Although such patterns had already been encountered in the paintings of the Alhambra Palace in Granada (see Chapter 10), crystallographers generally favour the Penrose tiling. Some examples are presented in Fig. 7.1.2 (Note 7.3).

Fig. 7.1.1 Examples of plane tilings found by Kepler (*Harmonice Mundi*, vol. 2, 1619).

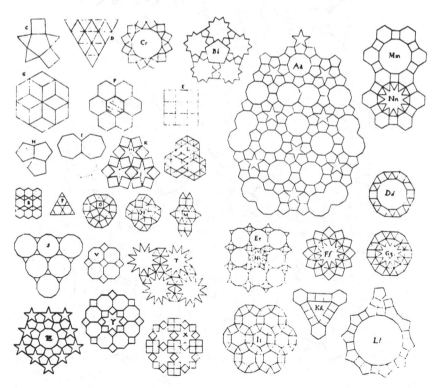

Fig. 7.1.2 Variants of the Penrose tiling: (*a*) tiling consisting of two rhombi; (*b*) tiling consisting of four elements.

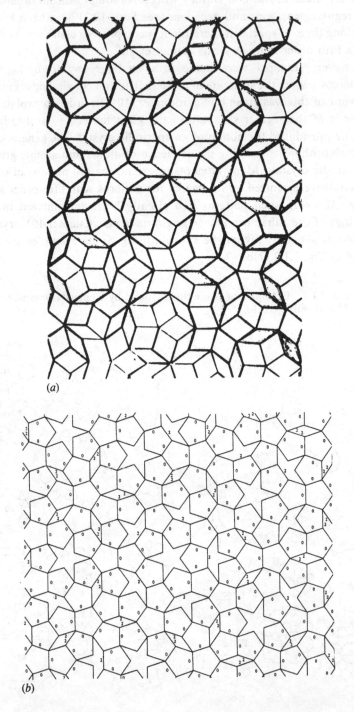

(*a*)

(*b*)

On the face of it, it is difficult to see any relation between the Penrose tiling and dynamic systems. However, this relation is demonstrated in Fig. 7.1.3 [20]. This figure shows how the Penrose tiling develops on the frame of the stochastic web already presented in Fig. 6.4.4a. The set of

Fig. 7.1.3   The relationship between the Hamiltonian $H_{q=5}$ and the Penrose tiling: (*a*) regions corresponding to the values of function H_5 lying inside the interval $H_5 \in (-0.85; 1.15)$; (*b*) the technique of constructing the Penrose tiling by connecting certain points of the relief in (*a*); the algorithm of connection is straightforward from the pictures.

(*a*)

(*b*)

singularities (elliptic and hyperbolic points, separatrices) of a dynamic system forms patterns in phase space. These patterns in certain cases turn out to be connected with patterns of the quasi-symmetric type. Due to this relation, some geometrical properties of patterns can be studied by dynamic methods, while certain properties of dynamic systems are revealed by structural methods.

7.2 Dynamic generation of patterns

One of the first methods of creating quasi-crystal patterns is based on the projection of N-dimensional densely packed cubes from an N-dimensional space onto a D-dimensional one ($D < N$). For $D = 2$, $N = 5$ and for a properly chosen angle for the plane of projection, the result was the Penrose tiling (Note 7.4).

The main idea of the projection method consists of the following. Consider a set of orthogonal vectors $\{e_j\}$ ($j = 1, \ldots, q$) issuing from a single point ('hedgehog' $\{e_j\}$). In Fig. 7.2.1 we see a two-dimensional ($N = 2$) and a three-dimensional 'hedgehog'. Let us correlate an infinite set of hyperplanes, $(q-1)$-dimensional in q-dimensional space, orthogonal to the vector, evenly spaced and parallel to each other, to each vector e_j. Let us call this family of hyperplanes a 'grid'. The set of q-grids comprises a multigrid of the order of q. The intersections of hyperplanes form a q-dimensional generalization of a square lattice. The projection of a certain portion of the multigrid, enclosed in a certain hyperlayer of finite thickness, onto D-dimensional space forms a tiling, or mosaic. In this way the Penrose tiling appears, and it can be proved that it does not contain any 'holes' or intersections of rhombi [22].

In a more general case, the organization of the grid may be more complex (for example, it might include some white spaces) but it is clear

Fig. 7.2.1 A two-dimensional (a) and three-dimensional (b) 'hedgehog'.

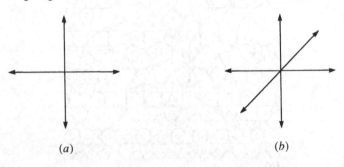

(a)   (b)

that projection is evidently a linear operation. This important conclusion means that regular multigrids, when projected, form a complex tiling of regular patterns possessing long range order. However, the long range order is connected not with the periodicity of patterns, but either with their quasi-periodicity (a finite number of incommensurate periods) or with their conditional periodicity (an infinite number of incommensurate periods).

There are further generalizations of the idea of how to use projections in creating aperiodic patterns. For example, a set of vectors $\{e_j\}$ can form an arbitrary (irregular) 'hedgehog' while the family of hyperplanes can be replaced by a family of hypersurfaces [17].

Let us go back to mapping $\hat{M}_q$ with the resonant twisting (see Sect. 6.1):

$$\hat{M}_q: \begin{cases} \bar{u} = (u + K_0 \sin v) \cos \dfrac{2\pi}{q} + v \sin \dfrac{2\pi}{q} \\[2mm] \bar{v} = -(u + K_0 \sin v) \sin \dfrac{2\pi}{q} + v \cos \dfrac{2\pi}{q}. \end{cases} \tag{7.2.1}$$

It operates on the plane (u, v) and has invariant elements s

$$\hat{M}_q s = s. \tag{7.2.2}$$

The set of elements s consists of sets of fixed points and stochasticity regions. For example, the stochastic web is one of such elements. Through the application of mapping $\hat{M}_q$, a point belonging to a stochastic web is mapped onto another point belonging to it. However, the web as a whole is invariant in respect to the action of $\hat{M}_q$ and, therefore, $\hat{M}_q$ might be considered as a generator of tilings with quasi-symmetry of the order of q. The word 'quasi-symmetry' here means that we are dealing with a tiling of the plane which is not exactly regular in the usual sense.

The important property of the definition of symmetry by means of dynamic operator $\hat{M}_q$ (7.2.2) is the fact that the invariant set of points $\{s_q\}$ comprising the stochastic web appears in the course of time. Only at $t \to \infty$ can the set of points $\{s_q\}$ adequately represent the web's structure.

The operation of smoothing the web and eliminating its minor details has a distinct dynamic meaning. Let $H(u, v, t)$ be the Hamiltonian of the initial system, the equations of motion for which are equivalent to mapping $\hat{M}_q$. The expression for $H(u, v, t)$ has been given in Sect. 6.1 (equation (6.1.15)):

$$H(u, v, t) = \frac{\pi}{q} (u^2 + v^2) - K_0 \sin v \sum_{n=-\infty}^{\infty} \delta(t - n) \tag{7.2.3}$$

where $\alpha = 2\pi/q$ is assumed and t substituted for τ. Equations of motion have the usual Hamiltonian form:

$$\dot{u} = \frac{\partial H}{\partial v}; \qquad \dot{v} = -\frac{\partial H}{\partial u}. \qquad (7.2.4)$$

The web's smoothing determined by the Hamiltonian (7.2.3), is done by a transition to a rotating coordinate system with period $t_q = q$ and the subsequent time averaging. This leads to a new Hamiltonian (see Sect. 6.4, equation (6.4.7)):

$$H_q = \sum_{j=1}^{q} \cos(\boldsymbol{\rho} e_j) \qquad (7.2.5)$$

where constant $(-\Omega_q/2)$ was omitted and the set of basis vectors e_j was redefined for the sake of convenience

$$\boldsymbol{\rho} = (v, u); \qquad e_j = \left(\cos\frac{2\pi}{q}j, \sin\frac{2\pi}{q}j \right). \qquad (7.2.6)$$

The new Hamiltonian (7.2.5) describes a new dynamic system with the following equations of motion:

$$\dot{u} = \frac{\partial H_q}{\partial v}; \qquad \dot{v} = -\frac{\partial H_q}{\partial u}. \qquad (7.2.7)$$

However, the system now has only one degree of freedom (H_q is time-independent) and, therefore, is formally integrable. Specifically, the set (7.2.7) is free of chaotic trajectories. The Hamiltonian H_q in (7.2.5) has been used for the explanation of fifth-order symmetry (in the case of $q = 5$) [16, 25]. It bears much interesting information, especially if we take into account the origin of the Hamiltonian H_q and its connection with mapping $\hat{M}_q$. Consider the surface $H_q = H_q(u, v)$. Its cross-section by the plane

$$H_q(u, v) = E \qquad (7.2.8)$$

defines the contours at that level. Since H_q is the Hamiltonian of the dynamic system (7.2.7) with energy E, lines of the level of equation (7.2.8) form a family of close-type invariant curves of diverse shapes and sizes. Among these are singular elements: separatrices, elliptic and hyperbolic points. Singular points are found from the following equations:

$$\left.\frac{\partial H_q}{\partial v}\right|_{H_q = E} = 0; \qquad \left.\frac{\partial H_q}{\partial u}\right|_{H_q = E} = 0. \qquad (7.2.9)$$

We have already solved (7.2.9) for $q = 4$ (Sect. 6.2) and $q = 3$ and 6 (Sect. 6.4). For all $q \in \{q_c\}$ there exists a single level E_h on which lie all the hyperbolic points and only two levels E_{e_1} and E_{e_2} on which lie all the elliptic points.

A specific feature of dynamic systems H_q forming a tiling with quasi-symmetry ($q \notin \{q_c\}$) on the phase plane, is energy level E distribution of singularities. Figure 7.2.2 shows the number of elliptic points ρ_e for various energy values in the case of symmetry of the fifth and seventh order (the normalization of ρ_e is arbitrary) [18]. These points, e.g., for the case of $q = 5$, are distributed in the vicinity of two values of energy: $E_{e_1} = -3$ and $E_{e_2} = 1.5$ which correspond to bottoms of potential wells and to peaks of potential humps. The distribution of ρ_e in the case of $q = 7$ is more even. The measure of uniformity increases with the growth of q.

A similar distribution of hyperbolic points ρ_h for $q = 5$ and $q = 7$ is presented in Fig. 7.2.3. Since a separatrix passes through each hyperbolic point, drastic changes of the phase portrait of a system must follow changes of energy E. We may say that systems with quasi-symmetry are densely filled by catastrophes in the energy interval $(-q, q)$. It becomes clear from Fig. 7.2.2 and Fig. 7.2.3 that values of energy E also exist for which the number of singular points has local maximums. The drastic difference between phase portraits at various E is well presented in Fig. 6.4.4*a*, *c* for $q = 5$.

Thus, unlike the usual periodic lattices where singularities of the energy surface are situated on planes of strictly fixed energy values, in our case of $q \notin \{q_c\}$ there exists a dispersion in the distribution of these points. This picture is characteristic of disordered systems. Therefore, systems (7.2.5) form a new and as yet poorly understood class of dynamic systems with a quasi-regular pattern of the phase portrait. Although systems H_q possess long range order, in some of their properties they approach irregular and disordered systems. All this is due to a high score of bifurcations of the phase portrait which is very sensitive to small energy variations.

Consider a small non-stationary perturbation of the Hamiltonian $H_q(u, v)$. All of its separatrices must be destroyed and dressed in stochastic layers. As shown, for example, by Fig. 6.4.4*c* for $q = 5$, separatrix loops may be situated far from each other. However, if they lie close to each other, as in Fig. 6.4.4*a* and *b*, their stochastic layers may merge. Narrow spacings between adjacent sections of non-disturbed separatrices are 'filled' and comprise a uniform network which may cover the whole plane. This is a stochastic web in the dynamic system (7.2.3) and in the

Fig. 7.2.2   The distribution of elliptic points for (a) $q = 5$ and (b) $q = 7$.

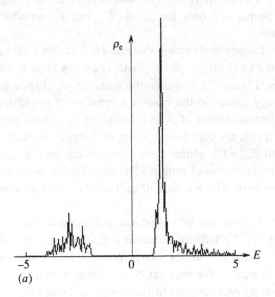

ρ_e

E

-5 0 5

(a)

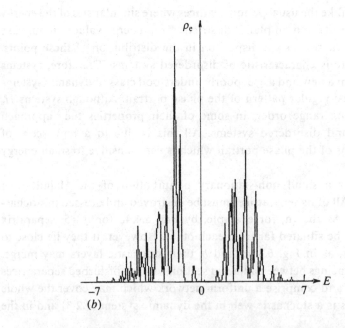

ρ_e

E

-7 0 7

(b)

Fig. 7.2.3 The distribution of hyperbolic points for (*a*) $q = 5$ and (*b*) $q = 7$.

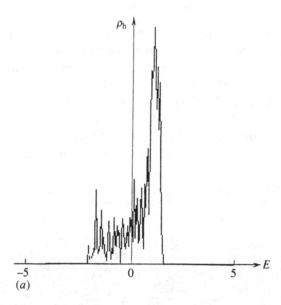

(*a*)

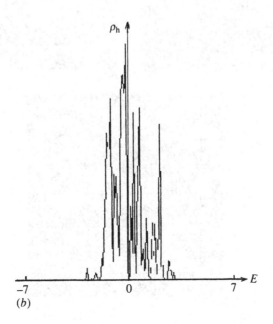

(*b*)

mapping (7.2.1) engendered by it. During the transition

$$H(u, v, t) \to H_q(u, v) \qquad (7.2.10)$$

the part of the Hamiltonian, or perturbation, which leads to the destruction of separatrices and formation of the web is lost.

For the averaged Hamiltonian $H_q(u, v)$, a web appears only in the vicinity of the energy layer with $E \approx E_c$. The value of E_c corresponds to the maximum of hyperbolic points in distribution $\rho_h(E)$. We see in Fig. 7.2.3 that, for $q = 5$, $E_c \approx 1$ and, for $q = 7$, $E_c \approx -1$.

The above prompts a way of getting the frame of a stochastic web directly from the expression for $H_q(u, v)$. Consider the narrow energy

Fig. 7.2.4 Structural skeletons: (*a*) $q = 7$; the size of the square = $64\pi \times 64\pi$; (*b*) $q = 8$; the size of the square = $48\pi \times 48\pi$; (*c*) $q = 12$; the size of the square = $56\pi \times 56\pi$.

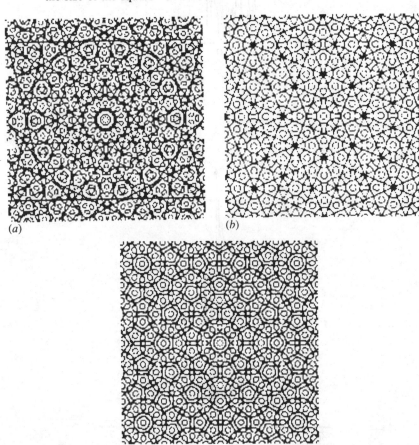

(*a*)

(*b*)

(*c*)

layer of width ΔE in the vicinity of E_c. A set of points belonging to the system and satisfying equation (7.2.8) and the following condition

$$E \in (E_c - \Delta E/2, E_c + \Delta E/2)$$

forms a tiling of the plane – namely, the skeleton of patterns. The examples of such tilings for $q = 5, 7, 8, 12$ are presented in Fig. 7.1.3*a* and Fig. 7.2.4. They contain the webs' skeletons. The smearing of all trajectories over a small region within the energy interval of the width of ΔE plays the role of perturbation.

By these means, with the help of transition (7.2.10) and the subsequent separation of the skeleton of patterns we establish the relation between the dynamical generator of tilings $\hat{M}_q$ and mosaics of a plane tiling (Note 7.5).

7.3 Quasi-symmetry, Fourier spectrum and local isomorphism

We have introduced a rather vague notion of quasi-symmetry in this chapter. Its intuitive introduction is closely related to fractal properties of chaos. Consider, for example, a stochastic web with fourth-order symmetry ($q = 4$). The smaller the parameter of perturbation K_0, the more the web resembles a square lattice. However, the geometrical properties of the web do not end here. The border between the region of stochastic dynamics and the region of regular motion has a very complex fractal shape. The greater the accuracy with which we wish to determine the shape of the border, the more complex its structure. This property is inherent in any stochastic layer (Note 7.6).

Therefore, when defining the geometrical shape of an invariant region caused by a real trajectory within a thin stochastic layer, one scarcely speaks of exact symmetry. The interaction between a certain part of motion with translational symmetry and a part of motion with rotational symmetry (or simply the interaction of symmetries) is bound to damage both symmetries even in the case of small values of the interaction constant K_0. However, symmetry defects may be small. Therefore, symmetry exists in a certain approximate sense which might even be difficult to define clearly. We understand intuitively that a certain degree of coarsening makes the web's shape more regular, more symmetrical. Thus sometimes it is better to give up any definition of pure symmetry, instead turning to quasi-symmetry. The existence of quasi-symmetry means that the property of symmetry holds true for an infinite geometrical object with a certain degree of accuracy. The essential thing is that this accuracy is locally uniform, i.e., it is independent of the location of a given element of the object.

For example, for a web this means the following. Let us consider a portion of the web inside a circle of a certain radius R. Ideal symmetry means that the geometrical shape of the web inside the circle must satisfy certain relations. In practice, this is not necessarily so. However, deviations from ideal symmetry do not exceed a certain value $\delta(R)$. The existence of a locally uniform accuracy of preservation of symmetry implies that the two following relations must be satisfied:

(1) for circles of different locations on the plane, there are different values of $\delta_i(R)$ satisfying the condition

$$\delta_i(R) \xrightarrow[R\to\infty]{} \delta_{i0}, \qquad (7.3.1)$$

where values of δ_{i0} are R-independent;

(2) for all i, the following condition must be satisfied:

$$\delta_{i0} \leqslant \delta_0. \qquad (7.3.2)$$

From the point of view of the concepts introduced, mapping $\hat{M}_q$ for small values of parameter K_0 can be viewed as a generator of tilings of a plane with qth-order quasi-symmetry of the 'quasi-crystal' type. Smoothed patterns produced by the Hamiltonian H_q (7.2.5) are more regular. In the transition from a web to smoothed reliefs, lines get somewhat smoothed and certain elements disappear. Therefore, we can assume that these reliefs, the skeletons of webs, as for example in Fig. 7.2.4, are certain decorations of the web.

In general, on the basis of a web or its relief, or skeleton, one can construct various plane tilings by making use of an additional algorithm of connecting various points of the basic picture. This procedure is quite naturally named decoration (Note 7.7).

We have already seen one such decoration which resulted in Penrose tiling (Sect. 7.1, Fig. 7.1.3). Similarly, one can obtain a mosaic of the seventh order (Fig. 7.3.1) as a decoration of the skeleton in Fig. 7.2.4a. To create it, one needs only three types of rhombi with the following acute angles: $(\pi/7, 2\pi/7, 3\pi/7)$. This is the minimal number of rhombi or any other elements to form a tiling. A more complex type of tiling with symmetry of the seventh order is shown in Fig. 7.3.2. Any tiling obtained as a result of the decoration of a web can be transformed into another tiling by means of another decoration. Thus, an infinite number of mosaics with one and the same symmetry can be obtained [31].

In smoothed patterns in Fig. 7.1.3*a* and Fig. 7.2.4 we can distinguish a set of straight lines comprising a multigrid, i.e., a system of parallel straight lines, tilted *q* times by the angle $2\pi/q$. Consider, for example, a pentagrid in Fig. 7.1.3*a*. The set of straight lines in it has a width $\Delta\varepsilon$ while the distance between any given pair of adjacent lines is

$$a_n = a_0\tau_0^n$$

where a_0 is the minimal distance, τ_0 is the golden mean:

$$\tau_0 = (1+\sqrt{5})/2 = 2\cos(\pi/5),$$

while *n* is an integer. The multigrid is another decoration of the web.

Let us define the process of filtration in the following way. Let us, for example, delete in Fig. 7.1.3*a* all lines with the exception of those comprising a pentagrid with only two possible spacings between adjacent

Fig. 7.3.1 The mosaic of the seventh order which is the generalization of the Penrose tiling to the case of symmetry of the seventh order and is based only on three types of rhombi (Chernikov *et al.* [18]).

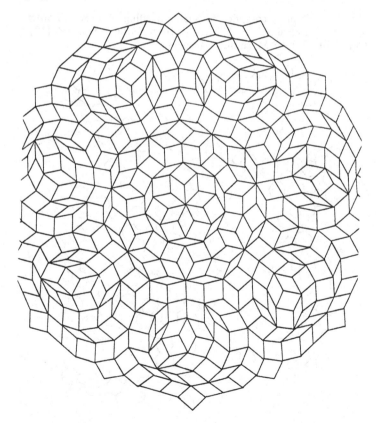

parallel lines: a_{n_0} and a_{n_0-1} ($n_0 \geq 1$). The pentagrid formed in this way is called the Ammann's lattice [4]. Coordinates x_m of straight lines in the Ammann's lattice satisfy the following simple rule [32]:

$$x_m = m + \beta_1 + \frac{1}{\tau_0}\left[\frac{m}{\tau_0} + \beta_2\right], \qquad (7.3.3)$$

where m is an integer; β_1, β_2 are constants and brackets $[\cdots]$ denote the integer part of the number.

Another way to map the sequence of alternations of two distances between the lines in the Ammann's lattice a_0 and b_0 has to do with the use of the following mapping [33]:

$$\hat{T}_s = \begin{pmatrix} 1 & 1 \\ 1 & 0 \end{pmatrix} \qquad (7.3.4)$$

applied to column-vector (a_0, b_0). It yields

$$a_1 = a_0 + b_0; \qquad b_1 = a_0;$$
$$a_2 = a_1 + b_1 = a_0 + b_0 + a_0; \qquad b_2 = a_1 = a_0 + b_0.$$

Fig. 7.3.2 A more complicated version of tiling with seventh-order symmetry. Reprinted by permission from *Nature*, **326**, 556 [19]. Copyright © 1987 Macmillan Magazines Ltd.

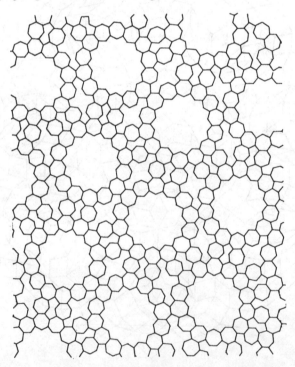

The sequence $\lim_{n\to\infty} a_n$ where the order of sequence of the distances a_0, b_0 is fixed, is called the Fibonacci set. It is this succession that determines the Ammann's lattice (7.3.3) in the case of $q = 5$.

In the case of $q = 7$, operator (7.3.4) must be substituted by another operator $\hat{T}_7$. It is nonlinear and the sequence of lines in the lattice is determined not by a single number, as in the case of $q = 5$, but by two numbers, e.g., $\tau_1 = \cos(\pi/7)$; $\tau_2 = \cos(2\pi/7)$ (Note 7.8).

The Ammann's lattice explains the essence of quasi-crystal patterns in the simplest possible way. The sequence of x_m in (7.3.3) is not a periodic one. However, it is quasi-periodic and contains two incommensurate periods (e.g., 1 and τ_0).

Thus, the stochastic web turns out to be connected with a certain class of aperiodic mosaics possessing the same symmetry as quasi-crystals. When $q \notin \{q_c\}$ the translational symmetry is lost (this is reflected in the fact that the mosaics are aperiodic). The property of translational symmetry is replaced by the property of local isomorphism which is defined in the following way. Let the tiling of a plane (mosaic) be comprised of a finite number of certain elements. For example, in the case of the Penrose tiling, such elements can be rhombi with the acute angles $\pi/5$ and $2\pi/5$. The number of various tilings is infinite. Let us consider one of them. If we distinguish a certain region, consisting of a finite number of elements, then the same region (or, more precisely, a congruent one) will be encountered an infinite number of times in this particular tiling, as well as in any other tiling consisting of the same elements (Note 7.9).

Local isomorphism of patterns with quasi-symmetry shows itself most vividly in their Fourier spectra. They correspond to the X-ray analysis of real crystals which enables us to form an opinion on the symmetry of the inner order of atoms and molecules in condensed matter (Note 7.10).

Let us denote by S_Γ a set of points belonging to a certain pattern on the phase plane of size Γ. Let us also assume that

$$\delta(\boldsymbol{R} - \boldsymbol{R}_a) = \begin{cases} 1, & a \in S_\Gamma \\ 0, & a \notin S_\Gamma, \end{cases} \tag{7.3.5}$$

where $\boldsymbol{R}$ is the vector of an arbitrary point on the plane Γ and $\boldsymbol{R}_a$ is the vector of the fixed point a. Then the Fourier image of the pattern is defined by the following expression:

$$S(k) = \lim_{\Gamma \to \infty} \frac{1}{(2\pi)^2} \int d\boldsymbol{R} \, e^{ik\boldsymbol{R}} \delta(\boldsymbol{R} - \boldsymbol{R}_a). \tag{7.3.6}$$

In practice, we are always dealing with a finite region Γ. This causes some additional boundary effects influencing the form of functions $S(k)$.

In the case of periodic patterns they can sometimes be easily discerned. However, in the case of aperiodic plane tilings, it is impossible to discern a pure 'monocrystal'. Nevertheless, when the region Γ is sufficiently large, Fourier spectra of various 'samples' Γ_j, slightly differing in shape and in size, also show only minor differences.

Figures 7.3.3 and 7.3.4 present Fourier spectra for a stochastic web caused by generator $\hat{M}_5$ and for the skeleton of the web caused by lines of the level of the Hamiltonian H_5, respectively. The region Γ was chosen in the form of a circle. The Fourier spectra in Figs. 7.3.3b and 7.3.4b closely resemble each other. According to this very important result we

Fig. 7.3.3   (a) an element of the web with $q = 5$ if the radius of the circle is 50π and (b) its Fourier spectrum.

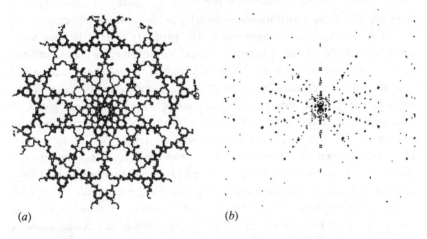

(a)   (b)

Fig. 7.3.4   (a) An element of the web's skeleton for $q = 5$ if the radius of the circle is 32π and (b) its Fourier spectrum.

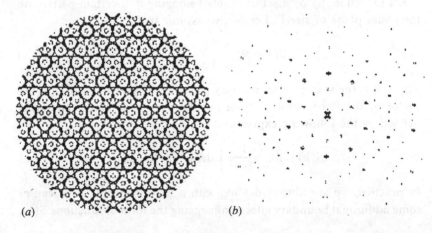

(a)   (b)

can analyse the symmetry of infinite webs in dynamic systems, in the same way as we do for crystals and quasi-crystals. A similar equivalence is shown in Figs. 7.3.5 and 7.3.6 for $q = 7$.

The process of local isomorphism can be most vividly demonstrated with the help of Fourier analysis. Consider, for example, a round sample of a web's frame with $q = 11$ (Fig. 7.3.7a). Its Fourier spectrum presented in Fig. 7.3.7b demonstrates that we are, in fact, dealing with a regular pattern with 11th-order symmetry (the point spectrum is lying on 22 rays issuing from a single centre). Let us extend the web's skeleton beyond the circle on the sample in Fig. 7.3.7a and choose a new sample of the

Fig. 7.3.5   (*a*) An element of the web at $q = 7$ if the radius of the circle is 24π and (*b*) its Fourier spectrum.

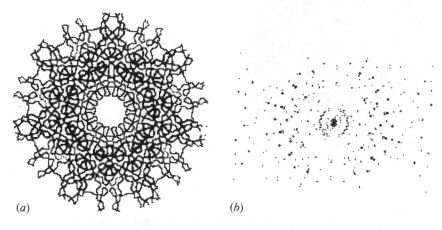

(*a*)   (*b*)

Fig. 7.3.6   (*a*) An element of the web's skeleton at $q = 7$ if the radius of the circle is 32π and (*b*) its Fourier spectrum.

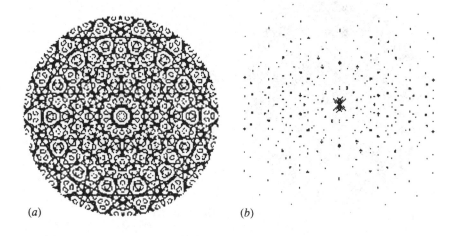

(*a*)   (*b*)

same shape and size but this time situated in a different portion of the plane. The new sample is presented in Fig. 7.3.8*a*. It can hardly be said that it has symmetry of the 11th-order. Rather, it resembles a texture. However, its Fourier spectrum in Fig. 7.3.8*b* definitely shows what object we are dealing with. Moreover, the comparison between spectra in Fig. 7.3.7*b* and 7.3.8*b* shows that they are practically indistinguishable.

Fig. 7.3.7 (*a*) An example of a web's skeleton with 11th-order symmetry chosen near the centre of symmetry and (*b*) its Fourier spectrum.

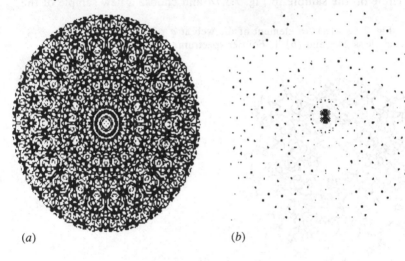

(*a*)   (*b*)

Fig. 7.3.8   As in Fig. 7.3.7, but the fragment is chosen in a different part of the plane, far from the centre of symmetry.

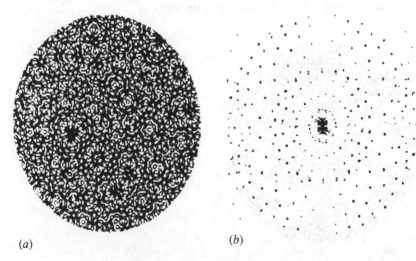

(*a*)   (*b*)

The above Fourier spectra of patterns are point ones (accurate to the error in their evaluation). This means that they describe regular patterns which cannot be classified as amorphous or disordered. But there are unexpected surprises awaiting us.

7.4 Singularities of the phase volume–energy dependency (Van Hove singularities)

The Hamiltonian of smoothed patterns (7.2.5):

$$H_q = \sum_{j=1}^{q} \cos(\rho e_j) \qquad (7.4.1)$$

makes it possible to analyse several subtle properties of the stochastic web. To these belong the Van Hove singularities related to the existence of elliptic and hyperbolic singularities in phase space [36]. Consider, for example, a system with one degree of freedom executing a finite motion with energy E. The phase volume $\Gamma(E)$ bounded by the following hypersurface:

$$H(p, x) = E \qquad (7.4.2)$$

(in our case it is a curve) is equal to

$$\Gamma(E) \doteq \int_{H \leq E} dp\, dx = \oint dx\, p(E, x), \qquad (7.4.3)$$

where integration in the last integral is performed over the interval of x-values which the curve $p = p(E, x)$ runs on hypersurface (7.4.2). The number of quantum states of a system with energy $\leq E$, as we know, is directly proportional to the phase volume $\Gamma(E)$, while the density of states is

$$\rho(E) = \text{const}\, \frac{d\Gamma(E)}{dE} = \text{const} \oint dx\, \frac{\partial p(E, x)}{\partial E}, \qquad (7.4.4)$$

where expression (7.4.3) has been used.

In the case of one degree of freedom, the integral on the right-hand side of (7.4.4)

$$\rho(E) = \text{const} \cdot T(E) \qquad (7.4.5)$$

coincides with the particle oscillation period $T(E)$ for energy E and singular points of the quantum density of states coincide with the singular points of the classical period of oscillation. For one degree of freedom, near an elliptic point there exists a discontinuity of the type of function jump caused by the border of admissible energy values. Near a hyperbolic point there exists a logarithmic singularity connected with the divergence of the period of oscillations on the separatrix.

In the case of dynamic systems defined by the Hamiltonian (7.4.1), motion is more complex (see the discussion in Sect. 7.2). To hypersurface (7.4.2) there corresponds not just one closed loop, but an infinite number of closed loops lying on the plane $H = E$ and forming a corresponding pattern on the phase plane. Therefore, formula (7.4.3) for the phase volume changes in the following way:

$$\Gamma(E) = \sum \oint dx\, p(E, x), \qquad (7.4.6)$$

where summation is performed over all the closed loops. Expression (7.4.6) diverges. However, there is another presentation for the density of states which does make sense. Consider the rectangle ($|p| \leq p_0$; $|x| \leq x_0$). Instead of (7.4.4) let us define the following expression:

$$\rho_0(E) = \lim_{p_0, x_0 \to \infty} \frac{1}{\Gamma_0} \sum \oint dx \frac{\partial p(E, x)}{\partial E}, \qquad (7.4.7)$$

where $\Gamma_0 = p_0 x_0$ and ρ_0 is the normalized density of states. The similarity property of the phase portrait defined by the Hamiltonian H_q for arbitrary q, ensures the existence of the limit in (7.4.7). Apparently, the property of local isomorphism also would suffice.

In the cases of $q = 3$, 4, 6 (symmetry of crystals), a single cell of the pattern will be enough to calculate ρ_0, since patterns are periodic. However, in the case of quasi-crystal symmetry, ($q \notin \{q_c\}$), the remark about the possibility of using equation (7.4.7) becomes non-trivial. Let us first give a few simple analytical expressions for $q = 4$ and $q = 3$.

For $q = 4$ let us write

$$H = \cos p + \cos x. \qquad (7.4.8)$$

Hence, equation (7.4.7) acquires the following form:

$$\rho_0(E) = \oint \frac{dx}{[1 - (E - \cos x)^2]^{1/2}} = 4K(1 - \tfrac{1}{4}\varepsilon^2)^{1/2} \qquad (7.4.9)$$

where K is the full elliptic integral of the first order. For $|E| \to 0$ and $|E| \to 2$, i.e., in the vicinity of a hyperbolic and an elliptic point, respectively, we have from (7.4.9)

$$\rho_0(E) = \begin{cases} 4\ln(8/|E|), & |E| \to 0 \\ 2\pi, & |E| \to 2. \end{cases} \qquad (7.4.10)$$

Numerical evaluation of $\rho_0(E)$ through the direct use of equation (7.4.7) is presented in Fig. 7.4.1.

For $q = 3$, let us write H as follows

$$H = \cos x + \cos\left(\frac{x}{2} + \frac{\sqrt{3}\,p}{2}\right) + \cos\left(\frac{x}{2} - \frac{\sqrt{3}\,p}{2}\right). \tag{7.4.11}$$

From (7.4.7) and (7.4.11) it follows that

$$\rho_0(E) = \frac{2}{\sqrt{3}} \oint dx\, [2(1 + \cos x) - (E - \cos x)^2]^{-1/2}. \tag{7.4.12}$$

Hence two energy intervals with different expressions for $\rho_0(E)$ follow. In the case of

$$-1 < E \leqslant 3,$$

we have the following from (7.4.12)

$$\rho_0(E) = \frac{8}{\sqrt{3}}(2E + 3)^{-1/4} K\left[\frac{\{[1 + (2\varepsilon + 3)^{1/2}]^2 - (1 + E)^2\}^{1/2}}{2(2E + 3)^{1/4}}\right].$$

Hence

$$\rho_0(E) = \begin{cases} 4\pi/3, & E \to 3 \\ 4\sqrt{3}\ln(1/|E + 1|), & E \to -1. \end{cases} \tag{7.4.13}$$

In the case of

$$-\tfrac{3}{2} \leqslant E < -1,$$

Fig. 7.4.1 The density of states ρ_0 and Van Hove singularities in the case of crystal symmetry: (a) $q = 3$; (b) $q = 4$. The absolute value of ρ_0 is in arbitrary units.

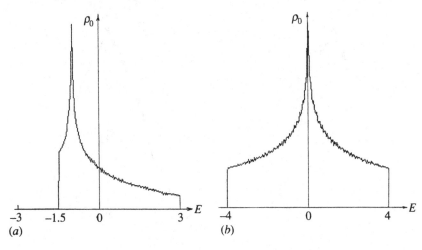

we have from (7.4.12)

$$\rho_0(E) = \frac{16}{\sqrt{3}} \{[1 + (2E+3)^{1/2}]^2 - (1+E)^2\}^{-1/2}$$

$$\times K\left[\frac{2(2E+3)^{1/4}}{\{[1+(2E+3)^{1/2}]^2 - (1+E)^2\}^{1/2}}\right].$$

Hence,

$$\rho_0(E) = \begin{cases} 16\pi/3 = 4\rho_0(3), & E \to -\frac{3}{2} \\ 4\sqrt{3}\ln(1/|E+1|), & E \to -1. \end{cases} \qquad (7.4.14)$$

Fig. 7.4.2   The density of state ρ_0 and Van Hove singularities in the case of quasi-crystal symmetry: (a) $q = 5$; (b) $q = 7$; (c) $q = 8$. The absolute value of ρ_0 is in arbitrary units.

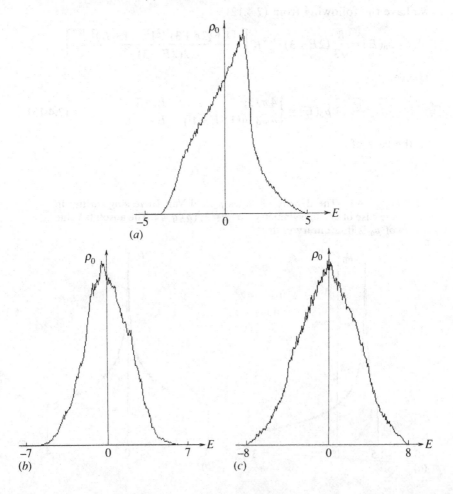

Values $E < -\frac{3}{2}$ for the Hamiltonian (7.4.11) are impossible. The corresponding form of $\rho_0(E)$, obtained numerically, is shown in Fig. 7.4.1.

All types of singularities defined by equations (7.4.10), (7.4.13) and (7.4.14) can be clearly seen in Fig. 7.4.1. Let us now turn our attention to the case of quasi-crystal symmetry. Numerical results for ρ_0 at $q = 5$, 7, 8 are presented in Fig. 7.4.2 [18]. This figure should be compared to Figs. 7.2.2 and 7.2.3 which show distributions of singular points of an elliptic (ρ_e) and hyperbolic (ρ_h) type over energy. The existence of their smeared distribution leads to smoothing out of Van Hove singularities. Visible singularities of ρ_0 are left only in such places where distributions of ρ_e and ρ_h have pronounced maxima. However, as the symmetry parameter q increases, the distribution of singular points of the Hamiltonian H_q becomes more regular and the density of states ρ_0 smoother. Beginning with $q = 7$, the function $\rho_0(E)$ is negligibly different from the cases of $q > 7$ and is closer to the density of states of a liquid than to that of a crystal. This property is very characteristic of quasi-crystal patterns. Despite the discrete (point) Fourier spectrum, some of their properties which are determined by the form of ρ_0 are analogous to the properties of a liquid.

7.5 Dynamic organization in phase space

In this chapter an important relation between the stochastic dynamics of a particle and the formation of regular patterns (i.e., patterns possessing a certain positive symmetry) has been established. Let us see in more detail how this happens. Consider the dynamics of a phase point on the phase plane, engendered by mapping $\hat{M}_q$ for small values of parameter K_0 and, for example, $q = 4$. Let us choose the initial condition so that it would fall within a stochastic web. Then the point performs a random walk on the phase plane. However, in the process of this random movement it creates an almost regular ornament with square-lattice symmetry. At first, some cells of this lattice are untouched. As the point continues to roam, the lattice becomes more and more regular. Thus, a well-organized figure appears asymptotically in the plane. This process might be called dynamic organization. It owes its existence to singular points and singular trajectories in the phase space of the system. This shows that chaos inherits structural properties of the phase space of the dynamic systems. In the case of weak chaos, structural properties manifest themselves more clearly (Note 7.11).

Now that we have learned of the organization of patterns in phase space, we feel the need to introduce certain criteria of organization. Until recently, our notion of the types of patterns was comparatively simple. It included crystals and liquids. Various amorphous, disordered media were usually classed together with liquids. The new notion of quasi-crystal symmetry called for more precise definitions. Or, rather, for a more definite understanding of what should be called a regular pattern and what a disordered or amorphous one. As we have seen in Sect. 7.3 and Sect. 7.4 of the present chapter, quasi-crystal patterns should be classified as structures with long range order, as evidenced, for example, by their Fourier spectra. However, the picture of the density of states $\rho_0(E)$ with smoothed out Van Hove singularities testifies to a disordered object rather than a crystal (Note 7.12).

To some extent, such paradoxes have to do with the fact that configurations of patterns in the case of quasi-crystal symmetry are described by a more complex algorithm than in the case of crystal order. For methods of analysis of patterns defined by complex algorithms, let us once more turn to the theory of dynamic systems [41].

Let vector ξ characterize the position of a certain element of the pattern while vector η defines the position of some other element. The relation between ξ and η can be specified by means of a certain operator $\hat{\mathcal{T}}(\eta \mid \xi)$:

$$\eta = \hat{\mathcal{T}}(\eta \mid \xi)\xi. \tag{7.5.1}$$

Let the points of the elements of the pattern lie on the trajectory of some dynamic system, as in the case of a dynamic web. Operator $\hat{\mathcal{T}}$ is induced by the dynamic operator $\hat{T}$ operating in phase space Γ of the initial dynamic system.

Equation (7.5.1) defines a new dynamic system in phase space of vectors (ξ). The number of elements of the pattern is denumerable. We juxtapose a vector to each element ξ_n. Then instead of (7.5.1) we can write:

$$\xi_{n+m} = \hat{\mathcal{T}}_m(n)\xi_n. \tag{7.5.2}$$

Let us introduce the correlator of two arbitrary integrable functions $f(\xi)$ and $g(\xi)$:

$$\mathcal{R}(f, g \mid m) = \int d\xi_n f(\xi_{n+m})g(\xi_n)$$

$$- \int d\xi_{n+m} f(\xi_{n+m}) \cdot \int d\xi_n g(\xi_n)$$

$$= \int d\xi f(\hat{\mathcal{T}}_m \xi)g(\xi) - \int d(\hat{\mathcal{T}}_m \xi)f(\hat{\mathcal{T}}_m \xi) \cdot \int d\xi g(\xi), \tag{7.5.3}$$

where definition (7.5.2) has been used.

With the help of (7.5.3) we can define the correlator at $g \equiv f$, as well as its Fourier expansion:

$$\mathscr{R}(f, f \mid \boldsymbol{\xi}) = \int_{-\infty}^{\infty} \mathrm{d}\boldsymbol{k} \, \mathrm{e}^{\mathrm{i}\boldsymbol{k}\boldsymbol{\xi}} \mathscr{R}(\boldsymbol{k}). \qquad (7.5.4)$$

It is natural to call value $\mathscr{R}(\boldsymbol{k})$ the pattern's spectrum. If mapping (7.5.2) defines a conditionally-periodic 'trajectory', its spectrum is discrete:

$$\mathscr{R}(\boldsymbol{k}) = \sum_{\nu} \mathscr{R}_{\nu} \delta(\boldsymbol{k} - \boldsymbol{k}_{\nu}) \qquad (7.5.5)$$

where a set of wave numbers $\boldsymbol{k}_{\nu}$ defines possible periods and orientations of characteristic planes of the pattern. For example, in the case of a periodic chain, there is only one value k_0 (with the exception of possible values nk_0). However, for the quasi-crystal Ammann's lattice the system of five oriented pairs of vectors $\boldsymbol{k}_1$ and $\boldsymbol{k}_2$ already emerges, the relation of their lengths being equal to the golden mean.

The other possibility has to do with the fact that spectrum $\mathscr{R}(\boldsymbol{k})$ may turn out to be continuous. This corresponds to the case of decay of correlations in (7.5.3) and disordered pattern. The degree of stochasticity of patterns depends upon how quickly the correlator (7.5.3) splits – according to the power or exponential law. Where the decomposition is exponential, the patterns correspond to turbulence in dynamic systems.

So far, we have discussed the organization of patterns in phase space of a dynamic system. In the following chapter we shall see that similar patterns may appear in some hydrodynamic problems in real space.

8
Two-dimensional hydro-dynamic patterns with symmetry and quasi-symmetry

We have shown that the phase portrait of a dynamic system with $1\frac{1}{2}$ degrees of freedom can have a certain not-too-complex periodical or almost periodical structure. Such properties of the system's phase plane require quite definite conditions, which imply the existence of a stochasticity region (a stochastic web). It was this web that caused a partitioning of the phase plane either with a periodic symmetry (or simply symmetry) or with a quasi-periodic symmetry (or quasi-symmetry). This led us to a new concept of symmetries produced by dynamic systems in phase space. To what extent are symmetries of partitioning of phase space universal and are similar symmetries found in other natural objects?

We have already mentioned quasi-crystals as an example of quasi-symmetrical patterns. In this chapter we are going to show that hydro-dynamic flows can also possess symmetry and quasi-symmetry similar to those displayed by the dynamic systems discussed earlier.

The formation of symmetry patterns in liquids has been known for a long time. An example of one of the simplest patterns (from the geometrical point of view) is Von Karman's vortex street [1]. It consists of two rows of evenly spaced point vortexes arranged as on a chess board. The model of the vortex street was proposed by Von Karman to explain periodic traces behind a streamline cylinder for Reynolds numbers $30 < \mathrm{Re} < 300$. Von Karman's street is an example of a one-dimensional periodic chain, and this structural property is widely used in the analysis of its characteristics [2, 3].

Among two-dimensional periodic patterns, the most famous is Benard's convective cell [4]. This cell is formed as a result of thermal convection (the so-called Rayleigh–Benard convection [4, 5]). In a liquid layer heated from below there appear ascending and descending streams. As a result of these convective motions, a hexagonal structure resembling a

Fig. 8.0.1 (a) The structure of Benard cells (according to Koschmieder and Biggerstaff [6a]); (b) The structure of square cells in the case of thermal convection (according to Le Gal *et al.* [6b]).

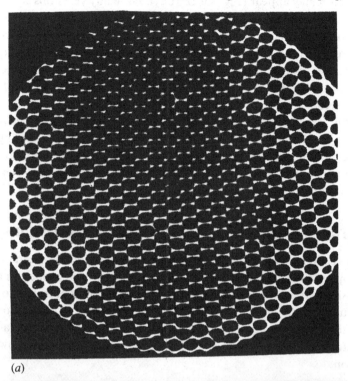

(a)

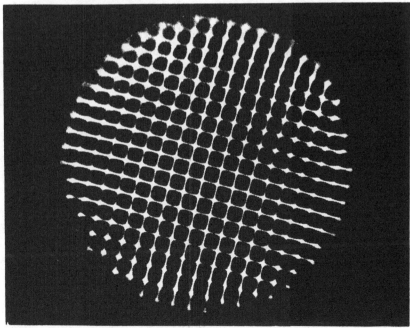

(b)

honeycomb may appear (Fig. 8.0.1*a* [6*a*]). For many years this structure has served as an example of a wonderful phenomenon of self-organization in continuous media. Depending upon specific conditions, various restructurings of the flow and, consequently, of the surface of the liquid are possible. Instead of hexagonal structures, structures with square cells may appear (Fig. 8.0.1*b* [6*b*]) as well as other still more complex configurations (Note 8.1).

As will become clear in the course of the present and the following chapters, in real coordinate space a stochastic web can exist serving as a route for diffusion of passive particles. The spatial web provides a certain partitioning of space. As a result, various portions of it possess specific properties. In this way, the pattern is formed in space.

8.1 Two-dimensional steady-state vortex flows in an ideal liquid

In this section we are going to dwell upon the description of certain of the simplest vortex flows of a liquid displaying structural properties and possessing certain symmetry properties. We shall restrict ourselves to the most simple situation of vortex motion, considering the plane flow of an ideal incompressible liquid in infinite space. The equation of conservation of matter for an incompressible liquid is reduced to the following equation:

$$\text{div } v = 0 \tag{8.1.1}$$

where $v(r, t)$ is the vector of the liquid's velocity, that is the main parameter of the flow, and a function of coordinates r and time t. The motion of an ideal incompressible liquid adheres to the Euler equation:

$$\frac{\partial v}{\partial t} + (v\nabla)v = -\nabla(P/\rho) \tag{8.1.2}$$

where P is the liquid's pressure and ρ is its density which is further assumed to be constant. The Euler equation (8.1.2) also allows for the following notation [10]:

$$\frac{\partial v}{\partial t} + \text{rot } v \times v = -\nabla(P/\rho + v^2/2). \tag{8.1.3}$$

If we apply the operation rot to both sides of equation (8.1.3), it is reduced to the following:

$$\frac{\partial \Omega}{\partial t} = \text{rot}[v \times \Omega] \tag{8.1.4}$$

where the value $\Omega = \mathrm{rot}\, v$ and is termed vorticity. Equation (8.1.3) is interpreted as the equation of frozen-in value Ω. For plane motions, when velocity $v = (v_x, v_y, 0)$ depends only upon two coordinates x and y, condition (8.1.1) is written in the following form

$$\frac{\partial v_x}{\partial x} + \frac{\partial v_y}{\partial y} = 0 \qquad (8.1.5)$$

which makes it possible to express velocity through the scalar function $\Psi(x, y, t)$ that is termed the stream function:

$$v_x = -\frac{\partial \Psi}{\partial y}; \qquad v_y = \frac{\partial \Psi}{\partial x}. \qquad (8.1.6)$$

For such two-dimensional flows, vorticity possesses only one z-component $\Omega = (0, 0, \Omega)$ and equation (8.1.4) is reduced to the following scalar equation

$$\frac{\partial \Omega}{\partial t} + (v\nabla)\Omega = 0; \qquad \Omega = \nabla^2 \Psi. \qquad (8.1.7)$$

It reflects the conservation of vorticity in its transport by liquid particles. The expressions for the liquid's velocity (8.1.6) and vorticity in terms of the stream function Ψ enable us to present equation (8.1.7) in the following form

$$\frac{\partial}{\partial t} \nabla^2 \Psi + J(\Psi, \nabla^2 \Psi) = 0 \qquad (8.1.8)$$

where a brief notation for the Jacobian has been assumed:

$$J(\Psi, \nabla^2 \Psi) = \frac{\partial \Psi}{\partial x} \frac{\partial}{\partial y} \nabla^2 \Psi - \frac{\partial \Psi}{\partial y} \frac{\partial}{\partial x} \nabla^2 \Psi. \qquad (8.1.9)$$

Let us consider steady-state vortex flows in an ideal liquid. In this case, the Jacobian (8.1.9) is zero. Hence it follows that vorticity Ω is constant along lines of the level of the stream function Ψ. The study of plane steady-state vortex flows in an ideal incompressible liquid is, consequently, reduced to finding the solution of the following equation [11]

$$\nabla^2 \Psi = f(\Psi), \qquad (8.1.10)$$

where f is an arbitrary function. This equation becomes linear if $f(\Psi)$ is either a constant or a linear function of Ψ and the general solution can be given for it. If, on the contrary, vorticity is a nonlinear function of Ψ, equation (8.1.10) becomes more complex. However, quite a few special solutions have been found for this case. (Note 8.2).

Now let us dwell upon a certain class of solutions of equation (8.1.10) which will be of special interest for us later as an example of a flow with a very distinct cellular structure and non-trivial symmetry properties. If $f = -\Psi$, the equation for Ψ becomes linear:

$$\nabla^2 \Psi + \Psi = 0 \tag{8.1.11}$$

and has the exact solution

$$\Psi = \sum_{j=1}^{q} C_j \cos(\boldsymbol{Re}_j + \alpha_j) \tag{8.1.12}$$

where vector $\boldsymbol{R} = (x, y)$, the set $\{e_j\}$ is an arbitrary set of q unity vectors and C_j and α_j are arbitrary constants. Specifically, if all amplitudes are equal, $C_j = \Psi_0/2$, all phases $\alpha_j = 0$, and if set $\{e_j\}$ is comprised by unit vectors forming a regular star

$$e_j = \left\{ \cos \frac{2\pi}{q} j, \sin \frac{2\pi}{q} j \right\}, \tag{8.1.13}$$

then this class of solutions with the following stream function

$$\Psi = \tfrac{1}{2}\Psi_0 \sum_{j=1}^{q} \cos\left[x \cos\left(\frac{2\pi}{q}j\right) + y \sin\left(\frac{2\pi}{q}j\right) \right] \tag{8.1.14}$$

possesses remarkable symmetry properties, already encountered in Chapter 7.

For $q = 2$ we have the periodic Kolmogorov flow with the stream function

$$\Psi = \Psi_0 \cos y. \tag{8.1.15}$$

For $q = 4$ the following stream function

$$\Psi = \Psi_0(\cos x + \cos y) \tag{8.1.16}$$

describes a cellular structure with square cells, while at $q = 3$ and $q = 6$ the following stream function:

$$\Psi = \Psi_0\left[\cos x + \cos\left(\frac{x}{2} + \frac{\sqrt{3}}{2} y\right) + \cos\left(\frac{x}{2} - \frac{\sqrt{3}}{2} y\right) \right] \tag{8.1.17}$$

describes a cellular structure with hexagonal cells. The same function describes the flow of liquid in the case of thermal convection producing Benard's cells. Let us note that, at $q = 2, 3, 4, 6$, symmetry of the flows with stream functions (8.1.15)–(8.1.17) is comparatively simple and coincides with the symmetry of two-dimensional crystals. Such flows possess both the translational and orientational invariance with respect to rotation at the angle $2\pi/q$ (for even q) or π/q (for uneven q). If the set of vectors $\{e_j\}$ does not form a regular star, the translational invariance may survive, the cells being rhombi.

New types of structures appear in the cases of $q = 5, 7, 8, \ldots$. These patterns are not periodic. Their exact analogue are quasi-crystals. We shall dwell upon these flows later (Note 8.3).

Now let us consider the case of the nonlinear relationship between vorticity and stream function. The so-called Stuart's solution [14] corresponds to the choice of function f in the form of the following exponent

$$\nabla^2 \Psi = e^{-2\Psi}. \tag{8.1.18}$$

Equation (8.1.18) has an exact solution

$$\Psi = \ln(C \cosh y + (C^2 - 1)^{1/2} \cos x); \qquad C \geqslant 1, \tag{8.1.19}$$

which describes a stationary pattern in the form of a street of distributed vortexes arranged periodically along the x-axis with spacings between them equal to 2π. Parameter C characterizes the density of vorticity. For $C \to \infty$, the stream function (8.1.19) has the following form:

$$\Psi = \ln(\cosh y + \cos x) + \text{const} \tag{8.1.20}$$

coinciding with the stream function of a chain of point vortexes arranged periodically along the x-axis [2, 3].

Various patterns in flows can be visualized with the help of stream lines. Stream lines are lines in any point of which velocity is directed tangent to them. For a stationary field of velocity, stream lines coincide with the trajectories of liquid particles and become visible if one strews or scatters some passive particles in the form of a light and easily noticeable powder over the liquid's surface. When photographed with a short exposure time, the surface displays well-defined lines coinciding with the stream lines.

Equations of stream lines in the plane case are defined by the following set of differential equations:

$$\frac{dx}{v_x} = \frac{dy}{v_y}. \tag{8.1.21}$$

Substituting (8.1.6) into this we get:

$$\frac{\partial \Psi}{\partial x} dx + \frac{\partial \Psi}{\partial y} dy = d\Psi = 0 \tag{8.1.22}$$

from whence, $\Psi(x, y) = \text{const}$. By these means, in two-dimensional steady-state flows, stream lines are shown to have a very simple form and are identical with the family of lines of the level of function $\Psi(x, y)$.

Generally, only one stream line can be drawn through each point on a plane. The exceptions to this rule are such singular points through which several stream lines pass. These singular points correspond to saddle points on the plane $h = \Psi(x, y)$ and are connected by special separatrix stream lines.

In the case of flows possessing crystal symmetry $q = 3, 4, 6$, singular points lie on the same line of the level of stream function Ψ and are connected by a single separatrix network (see Sect. 7.4).

Figure 6.4.1 shows lines of the level of the stream function (8.1.16) corresponding to a flow with square cells. The surface $h = \Psi(x, y)$ at $q = 4$ looks like a chequerwork of peaks $\Psi > 0$ and valleys $\Psi < 0$. Separatrix stream lines form a square lattice at $\Psi = 0$. Figure 6.4.2 illustrates the lines of the level of the stream function (8.1.17) for $q = 3$. The cross-section of the surface at $\Psi = -\Psi_0$ forms kagome lattice separatrices. For other values of the level, stream lines are closed-type.

The picture of a flow with quasi-symmetry is different. In this case, saddle points of the surface $h = \Psi(x, y)$ are situated on different lines of the level and are not connected by a single separatrix net. Figure 6.4.4 presents the surface $h = \Psi(x, y)$ for $q = 5$. In it, we clearly see that saddle points lie at different heights. Nevertheless, saddle regions form patterns comprised of almost straight lines. On different planes $\Psi(x, y) = \text{const}$, there develop different distributions of separatrix loops, and singular stream lines do not form a global web-like network. There are numerous ruptures as can be seen in Fig. 6.4.4b, c displaying lines of the level of quasi-symmetry $q = 5$. The distance between lines of the level in the vicinity of ruptures at a certain level is very small. This has to do with the existence of a singularity in the density of saddle points distribution with height h at this level.

Similarly, the picture of stream lines for the flow (8.1.19) presented in Fig. 8.1.1 gives a graphic image of the vortex street. Such a flow possesses symmetry with respect to a shift by the period along the chain's axis.

The plane of a constant value of stream function and lines of the level play the same role as the phase plane of a dynamic system and its orbit. This similarity is most apparent if we take into account the fact that the equations of the trajectories of the liquid's particles, in the case of a plane motion, have the following form

$$\frac{dx}{dt} = -\frac{\partial \Psi}{\partial y}; \quad \frac{dy}{dt} = \frac{\partial \Psi}{\partial x}, \tag{8.1.23}$$

i.e., they are written in the form of Hamiltonian equations where the part of the Hamiltonian is played by the stream function $\Psi(x, y, t)$, coordinates (x, y) acting as the canonic pair. Having written down the equations of trajectories in the Hamiltonian form, we arrive at the following conclusion. If the plane flow is stationary, i.e., $\Psi = \Psi(x, y)$, then equations (8.1.23) correspond to a conservative system with one degree of freedom and, hence, the problem of finding the particles' trajectories in a liquid is integrable and trajectories of liquid particles coincide with lines of the level of the stream function. For non-steady-state two-dimensional flows in a liquid, where stream function Ψ is an explicit function of time, the set of equations (8.1.23) corresponds to the non-conservative Hamiltonian system and is, generally speaking, non-integrable.

8.2 Stability of steady-state plane flows with symmetrical structure

In the previous section it has been shown that steady-state flows with symmetry and quasi-symmetry can exist in a two-dimensional case. Their structure and its symmetry can be shown with the help of the picture of lines of the level. Finally, we arrive at a complete analogy to the analysis of an appropriate dynamic system and its phase portrait. However, this analogy contains an essential difference. The steady-state flow under consideration might turn out to be unstable, the further study of its structure becoming more or less meaningless.

Fig. 8.1.1   Stuart flow lines (8.1.19) at $C = 1.03$.

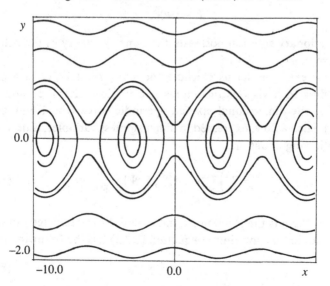

In this section we are going to broaden our notion of the existence of stationary structures and offer a few examples and arguments proving their ability to be stable. Moreover, patterns with, for example, quasi-symmetry can be forced when the system is under certain conditions.

Here we shall restrict ourselves to several of the simplest examples demonstrating the possibility of a stability region in flows with patterns determined by a stream function of the type of (8.1.14). As yet, the role of such flows possessing either translational symmetry or the symmetry of a quasi-crystal is not fully understood. This remark refers primarily to their stability in finite and infinite systems. Still, it is clear that under certain conditions in the form of an external field (forcing) and dissipation, such systems are possible.

First, let us consider the stability of a plane flow with the stream function (8.1.14) in an infinite system. With this aim in view, instead of equation (8.1.8) let us write the Navier–Stokes equation

$$\frac{\partial}{\partial t} \nabla^2 \Psi + J\{\Psi, \nabla^2 \Psi\} = \nu \nabla^4 \Psi + F, \qquad (8.2.1)$$

which differs from (8.1.8) by a viscous term on the right-hand side and the external force $F = F(x, y, z)$ that acts as a source of excitation of matter. The stationary solution $\bar{\Psi}(x, y)$ in the form of (8.1.14) for the non-viscous equation (8.1.8) will, according to (8.1.11), also be a stationary solution of the viscous equation (8.2.1) provided the following condition is satisfied:

$$F = -\nu \bar{\Psi}, \qquad (8.2.2)$$

i.e., the stationary solution possesses the same symmetry as the external force (pumping).

As a first example, let us consider the study of stability of a plane periodic flow in a viscous incompressible liquid (8.1.15), arising as a result of a one-dimensional spatially-periodic force (the Kolmogorov flow) [21]. Such a flow, expressed in dimensionless variables, is described by the following equation:

$$\left(\frac{\partial}{\partial t} - \frac{1}{\mathrm{Re}} \nabla^2\right) \nabla^2 \Psi + J\{\Psi, \nabla^2 \Psi\} = \frac{1}{\mathrm{Re}} \cos y \qquad (8.2.3)$$

where $\mathrm{Re} = \Psi_0/\nu$ is the Reynolds number and Ψ_0 is characteristic value of Ψ. The stationary solution of equation (8.2.3) has the following form:

$$\bar{\Psi} = -\cos y. \qquad (8.2.4)$$

For the perturbation $\Psi' = \Psi - \bar{\Psi}$, from (8.2.3) and (8.2.4) we have the following linearized equation:

$$\left(\frac{\partial}{\partial t} - \frac{1}{\mathrm{Re}} \nabla^2\right) \nabla^2 \Psi' - \sin y (1 + \nabla^2) \frac{\partial \Psi'}{\partial x} = 0. \qquad (8.2.5)$$

Let us restrict ourselves to the investigation of stability of perturbations periodic along the x-axis:

$$\Psi'(x, y, t) = e^{i\alpha x} \Psi'(y, t) + \text{c.c.} \qquad (8.2.6)$$

After substituting (8.2.6) into (8.2.5), we get, for perturbation $\Psi'(y, t)$, the so-called Orr-Sommerfeld equation:

$$\left(\frac{\partial}{\partial t} - \frac{1}{\mathrm{Re}} \nabla^2\right) \nabla^2 \Psi' + i\alpha \sin y (1 + \nabla^2) \Psi' = 0. \qquad (8.2.7)$$

Here $\nabla^2 = \partial^2/\partial y^2 - \alpha^2$. Let us replace perturbation $\Psi'(y, t)$ by its Fourier series in y:

$$\Psi'(y, t) = e^{\gamma t} \sum_{n=-\infty}^{\infty} \Psi_n e^{iny}. \qquad (8.2.8)$$

Substituting (8.2.8) into (8.1.7) we obtain an infinite set of equations for coefficients Ψ_n:

$$\frac{2}{\alpha} [\alpha^2 + n^2]\left(\gamma + \frac{\alpha^2 + n^2}{\mathrm{Re}}\right)\Psi_n$$
$$+ \Psi_{n-1}[\alpha^2 - 1 + (n-1)^2] - \Psi_{n+1}[\alpha^2 - 1 + (n+1)^2] = 0. \qquad (8.2.9)$$

It turns out that from the stability point of view the main risk is presented by long-wave perturbations ($\alpha \ll 1$). To study the stability with respect to such perturbations, it is sufficient to take into account only a finite number of harmonics with $n = 0, \pm 1$ in the system (8.2.9). Then as the solvability condition for the reduced set we obtain the following dispersion equation:

$$\left(\gamma + \frac{1 + \alpha^2}{\mathrm{Re}}\right)\left(\gamma + \frac{\alpha^2}{\mathrm{Re}}\right) - \frac{\alpha^2(1 - \alpha^2)}{2(1 + \alpha^2)} = 0. \qquad (8.2.10)$$

It follows from (8.2.10) that for sufficiently large Reynolds numbers (i.e., for $\mathrm{Re} \to \infty$) $\gamma^2 \approx \alpha^2/2 > 0$ and the Kolmogoroff flow is unstable with respect to long-wave perturbations. The critical value $\mathrm{Re} = \mathrm{Re}_c$ at which instability can disappear is found by setting $\gamma = 0$ in (8.2.10). Then for $\alpha \ll 1$, the critical value of the Reynolds number is given by the following asymptotic expression:

$$\mathrm{Re}_c = \sqrt{2}\,(1 + 3\alpha^2/2) + O(\alpha^4). \qquad (8.2.11)$$

For small overcriticality $\mathrm{Re} - \mathrm{Re_c} \ll \mathrm{Re_c}$, the expression for the growth rate of instability proves to be proportional to the square of the wave number of the long-wave perturbations

$$\gamma = \alpha^2(\mathrm{Re} - \mathrm{Re_c}). \tag{8.2.12}$$

Such growth rate-wave number relationship can be interpreted as an instability of the flow with negative viscosity (Note 8.4).

As a second example, let us consider the stability of the flow (8.1.16) with square cells. The equation describing this flow differs from (8.2.3) on the right-hand side

$$\left(\frac{\partial}{\partial t} - \frac{1}{\mathrm{Re}} \nabla^2\right) \nabla^2 \Psi + J\{\Psi, \nabla^2 \Psi\} = \frac{1}{\mathrm{Re}} (\cos x + \cos y), \tag{8.2.13}$$

while its stationary solution is the stream function

$$\bar{\Psi} = -(\cos x + \cos y). \tag{8.2.14}$$

The linearized equation has the following form:

$$\left(\frac{\partial}{\partial t} - \frac{1}{\mathrm{Re}} \nabla^2\right) \nabla^2 \Psi' - \sin y(1+\nabla^2)\frac{\partial \Psi'}{\partial x} + \sin x(1+\nabla^2)\frac{\partial \Psi'}{\partial y} = 0. \tag{8.2.15}$$

Presenting perturbation Ψ' in the form of a Fourier series

$$\Psi'(x, y, t) = e^{\gamma t + i\alpha x + i\beta y} \sum_{m,n} \Psi_{m,n}\, e^{imy + inx}, \tag{8.2.16}$$

we obtain for coefficients Ψ_{mn} an infinite set of equations

$$[(n+\alpha)^2 + (m+\beta)^2]\left[\gamma + \frac{(n+\alpha)^2 + (m+\beta)^2}{\mathrm{Re}}\right]\Psi_{m,n}$$

$$-\frac{(n+\alpha)}{2}\{[(n+\alpha)^2 + (m-1+\beta)^2 - 1]\Psi_{m-1,n}$$

$$-[(n+\alpha)^2 + (m+1+\beta)^2 - 1]\Psi_{m+1,n}\}$$

$$+\frac{(m+\beta)}{2}\{[(m+\beta)^2 + (n-1+\alpha)^2 - 1]\Psi_{m,n-1}$$

$$-[(m+\beta)^2 + (n+1+\alpha)^2 - 1]\Psi_{m,n+1}\} = 0. \tag{8.2.17}$$

As in the case of the Kolmogorov flow, it turns out that the most unstable are the long-wave perturbations, the wave vector for which is directed along one of the coordinate axes. In order to study the stability of these perturbations, let us, as in the previous example, take into account the finite number of harmonics with $m = 0, \pm1$; $n = 0, \pm1$ in expansion

(8.2.16). Considering, for simplicity's sake, $\beta = 0$ as the solvability condition for the truncated system, we get the following dispersion relation:

$$
[1+\alpha^2]\left(\gamma+\frac{1+\alpha^2}{\text{Re}}\right)
$$
$$
=\frac{\alpha^2(1-\alpha^2)}{2\left(\gamma+\dfrac{\alpha^2}{\text{Re}}\right)}-\frac{\alpha^2}{4}\left\{\frac{(\alpha-1)^2}{[1+(\alpha-1)^2]\left[\gamma+\dfrac{1+(\alpha-1)^2}{\text{Re}}\right]}\right.
$$
$$
\left.+\frac{(\alpha+1)^2}{[1+(\alpha+1)^2]\left[\gamma+\dfrac{1+(\alpha+1)^2}{\text{Re}}\right]}\right\}. \qquad (8.2.18)
$$

For $\alpha \ll 1$, an expression for the critical Reynolds number has the following form:

$$
\text{Re}_c = \sqrt{2}\,(1+13\alpha^2/8)+0(\alpha^4), \qquad (8.2.19)
$$

while, for the instability growth rate γ in the case of a small supercriticality, equation (8.2.12) holds true, i.e., we are again dealing with the same instability as for a flow with negative viscosity.

The above examples show that, at a sufficiently large viscosity $\text{Re} < \text{Re}_c$ and at a properly chosen forcing for the flows with patterns of the type of (8.1.14) for $q = 2$ and $q = 4$, there exists a stability region. In [29] a similar problem concerning the stability of structural flows with sixth-order symmetry ($q = 3$ and $q = 6$) was considered and a stability region has also been found.

The problem of the analytical study of stability of flows with quasi-symmetry is more complex, so that here we shall have to turn to a numerical experiment. A non-trivial question arises concerning the role of boundary conditions which, for example, being periodicity conditions, possess symmetry completely different from that of the forcing. In this case, a settled steady-state flow appears as a result of the competition between two different symmetries. Survival is possible for a structure possessing either one of these two symmetries or their complex combination.

Now let us show how a flow with quasi-symmetry appears. To be more specific, let us assume that the forcing in equation (8.2.1) has symmetry of the fifth order

$$
F(x,y) = F_0 \sum_{j=1}^{5} \cos\left\{2\pi k\left[x\cos\left(\frac{2\pi}{5}j\right)+y\sin\left(\frac{2\pi}{5}j\right)\right]\right\}. \qquad (8.2.20)
$$

The results of the numerical analysis of the problem for the grid of 128×128 meshes are shown in Fig. 8.2.1 [17]. These results reflect the case when boundary conditions are chosen in the following form:

$$\Psi(0, y; t) = \Psi(0, y; 0) = \Psi_0;$$

$$\Psi(1, y; t) = \Psi(1, y; 0) = \Psi_0;$$

$$\Psi(x, 0; t) = \Psi(x, 0; 0) = \Psi_0 \cos 2\pi x; \qquad (8.2.21)$$

$$\Psi(x, 1; t) = \Psi(x, 1; 0) = \Psi_0 \cos 2\pi x.$$

Such boundary conditions possess symmetry corresponding to $q = 2$. They determine a flow of liquid with the gradient of velocity profile

Fig. 8.2.1 Formation of a flow pattern with fifth-order symmetry. This is the result of the numerical integration of equation (8.2.1) with the external force given by (8.2.23). Parameters used: $k = 10$, $F_0 = 10^{-4}$, $\Psi_0 = 10^{-8}$, $\nu = 0.31$. Figures correspond to the following time instants: (a) $t = 0$; (b) $t = 2 \times 10^{-5}$; (c) $t > 10^{-2}$.

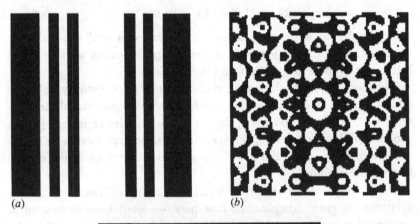

(a)   (b)

(c)

$v_y = -\partial \Psi/\partial x$ as a function of x, homogenous along the y-axis (see Fig. 8.2.1a). Thus, a competition occurs between the second-order symmetry imposed by the boundary conditions (8.2.21), and the fifth-order symmetry of the forcing (8.2.20). The resulting distribution of velocity is shown in Fig. 8.2.1c where parameters have been chosen so that the symmetry of the fifth order survives. Various intermediate cases are also possible. However, for further analysis, it is important to note that steady-state flows with the structure of quasi-crystals (among them, a pentagonal structure) are possible.

Now let us consider the situation where the forcing on the right-hand side of equation (8.2.1) is an explicit function of time.

To be more specific, let us choose the external force F in the following form:

$$F(x, y, t) = F_0(t) \cos[k(x \cos \omega t + y \sin \omega t)], \qquad (8.2.22)$$

where the wave number of the forcing $k = 2\pi/L$ and frequency $\omega = 2\pi/T$, F_0 being the amplitude of the external force which, generally speaking, is time-dependent.

Equation (8.2.1) with the external force (8.2.22) was numerically integrated on a grid of 128×128 meshes in the form of a unit square [18]. Figure 8.2.2 shows the result for the case where the amplitude of the external force $F_0(t)$ is a sequence of square pulses of short duration τ with the period of succession $T_0 = 2\pi/\Omega$. The duration of a single pulse in numerical computations [18] was considerably shorter than other time scales in the problem. As the result of numerical computations, it was found out that steady-state stationary patterns, provided the resonance conditions $\Omega = q\omega$ are satisfied, can have symmetry of either a crystal or a quasi-crystal type.

An approximate analytical expression for the stationary stream function $\bar{\Psi}$ can be obtained in the following way. Let us approximate the amplitude of forcing $F_0(t)$ by the following series of δ-functions

$$F(x, y, t) = F_0 \sum_{n=-\infty}^{\infty} \delta(t/T_0 - n) \cos[k(x \cos \omega t + y \sin \omega t)]. \qquad (8.2.23)$$

Making use of the presentation (6.4.4) we easily find

$$F = \bar{F} + \tilde{F};$$

$$\bar{F} = \frac{F_0}{q} \sum_{j=1}^{q} \cos \xi_j; \qquad (8.2.24)$$

$$\tilde{F} = \frac{2F_0}{q} \sum_{j=1}^{q} \cos \xi_j \sum_{m=1}^{\infty} \cos\left[\frac{2\pi m}{q}\left(\frac{t}{T_0} - j\right)\right],$$

Fig. 8.2.2 The result of numerical integration of equation (8.2.1) with the external force (8.2.23). Lines of the level $\Psi(x, y) = \text{const}$ of the flow function for the steady-state pattern of the flow at the following parameters: (a) $F_0 = 100$, $\nu = 0.01$, $T = 0.25 \times 10^{-3}$, $\Psi_0 = 10^{-2}$, $L = 10^{-1}$, $T = 10T_0$; (b) $F_0 = 100$, $\nu = 0.01$, $T = 0.16 \times 10^{-3}$, $\Psi_0 = 10^{-2}$, $L = 10^{-1}$, $T = 8T_0$.

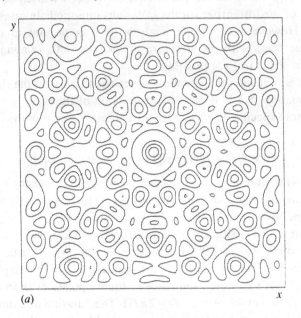

(a)

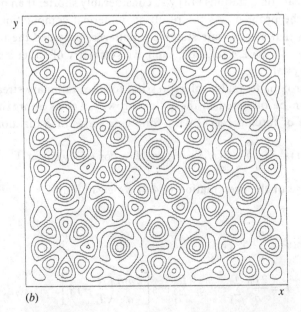

(b)

where

$$\xi_j = k\left[x \cos\left(\frac{2\pi}{q}j\right) + y \sin\left(\frac{2\pi}{q}j\right)\right].$$

The value $\xi_j = k\rho e_j$ where ρ is a radius-vector on the plane (x, y) and e_j is a unit vector defining the vertex of a regular q-square. Since function $\bar{F}(x, y)$ in (8.2.24) is the exact solution of equation (8.1.8), the stream function

$$\bar{\Psi} = -\frac{F_0}{\nu k^4 q} \sum_{j=1}^{q} \cos\left\{ k\left[x \cos\left(\frac{2\pi}{q}j\right) + y \sin\left(\frac{2\pi}{q}j\right)\right]\right\} \qquad (8.2.25)$$

is an exact solution of the Navier–Stokes equation (8.2.1) with the stationary forcing $\bar{F}$. Figure 8.2.2a shows results for the case when the resonance conditions $\Omega = 10\omega$ is satisfied, while Fig. 8.2.2b presents the set of lines of the level for the eighth-order resonance case.

Conditions on the boundary of a unit square Γ have been chosen as follows

$$\Psi(t, \Gamma) = \Psi_0, \qquad (8.2.26)$$

the initial conditions having been chosen in a similar way.

Figure 8.2.3 illustrates the solution of equation (8.2.1) where the amplitude of an external force has been chosen in the form of

$$F_0(t) = F_0 \cos \Omega t. \qquad (8.2.27)$$

The picture of the flow in the resonance case $\Omega = q\omega$ (q is an integer) differs qualitatively from the corresponding picture in a far-from-resonance case. In the resonance case, at $\omega/\nu k^2 \gg 1$ the final phase of the flow's evolution was a stationary structure with cylindrical symmetry of q-th order. Figure 8.2.3a displays the set of lines of the level of the stream function at an instant of time $t = 5T$ ($q = 10$) corresponding to the stationary pattern on the whole plane with the exception of a circle of radius $\sim L$ around the centre point, within which the flow continues to be non-stationary. In the case of reduced frequency ω or breaking of resonance conditions, there was no stationary pattern. In Fig. 8.2.3b a typical picture of lines of the level of the stream function at the stage of its evolution $t = 5T$ is shown. The value of mismatch of resonance condition $\delta\Omega = N\omega - \Omega$. In Fig. 8.2.3$b$ the relative value of mismatch of resonance $\delta\Omega/\Omega = 0.044$ and $\omega/\nu k^2 \cong 1$.

The numerical results presented above demonstrate the diversity of structures appearing in the simplest hydrodynamic flows. The mechanism behind various patterns has to do with complex interaction of rival forces

Fig. 8.2.3 The result of numerical integration of equation (8.2.1) with $F_0(t) = F_0 \cos \Omega t$ at (*a*) $F_0 = 0.03$, $\nu = 0.31$, $T = 0.25 \times 10^{-3}$, $\Psi_0 = 10^{-7}$, $L = 10^{-1}$; (*b*) $F_0 = 0.03$, $\nu = 0.31$, $T = 0.5 \times 10^{-2}$, $\Psi_0 = 10^{-7}$, $L = 10^{-1}$.

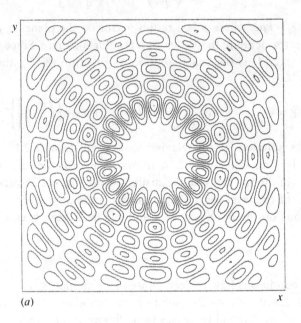

(*a*)

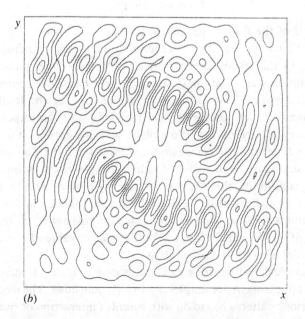

(*b*)

defining the dynamics of a fluid. In the general case, patterns can appear as manifestations of secondary flows or flows of a higher order. In such cases, the flows of a lower order play the role of force fields taking part in the competition and affecting the final form of the flow.

Now let us show how the structure of a surviving flow in the last example can be defined. Let us divide the external force (8.2.22) with amplitude (8.2.27) at $\Omega = q\omega$ into two parts - stationary $\bar{F}$ and nonstationary $\tilde{F}$.

$$F = \bar{F} + \tilde{F}; \qquad \bar{F} = F_0 J_q(k\rho) \cos q\varphi;$$

$$\tilde{F} \cong \frac{F_0}{2} \{ J_{q+1}(k\rho) \cos[(q+1)\varphi + \omega t] \tag{8.2.28}$$

$$+ J_{q-1}(k\rho) \cos[(q-1)\varphi + \omega t] \},$$

where ρ and φ are polar coordinates, $J_q(k\rho)$ is the Bessel function and on the right-hand side of the expression for $\tilde{F}$ the terms containing higher harmonics with frequency ω have been neglected. By virtue of the fact that function $\bar{F}$ is the exact solution of the two-dimensional Helmholz equation (8.1.11) expressed in polar coordinates, the stream function $\bar{\Psi} = C\bar{F}$ where C is an arbitrary constant makes the nonlinear term on the right-hand side of (8.2.1) equal to zero. Hence, the stream function

$$\bar{\Psi} = -\frac{F_0}{\nu k^4} J_q(k\rho) \cos q\varphi \tag{8.2.29}$$

is the exact solution of the Navier–Stokes equation with the stationary forcing $\bar{F}$. Provided $\omega \gg \nu k^2$, the quickly oscillating part of the stream function $\tilde{\Psi} \sim \tilde{F}/\omega k^2$ satisfies the condition $\tilde{\Psi} \ll \bar{\Psi}$. Therefore, a stationary pattern in the conditions of exact resonance presented in Fig. 8.2.3 is fairly well described by formula (8.2.29). The illustrations of two-dimensional flows presented above prove that it is possible to draw a certain analogy between the patterns emerging on the phase plane of a dynamic system and the structures of two-dimensional flows in an incompressible liquid. In actuality, this analogy can be extended much further. Here we must turn our attention to the analysis of the three-dimensional flows.

9
Chaos of stream lines

In the previous chapter we showed that in two-dimensional hydro-dynamics there can exist flows with patterns displaying symmetry and quasi-symmetry. However, the picture drawn in the previous chapter is poor compared to the one we are now going to present. Three-dimensional dynamics introduces us to a qualitatively new phenomenon – the existence of stream lines chaotically arranged in space – which is sometimes called the Lagrangian turbulence. Various forms of this phenomenon have interesting practical applications and have played an important role in our understanding of the onset of turbulence, as well. In this chapter we are going to establish the relation between the structural properties of steady-state three-dimensional flows and the chaos of stream lines in these flows. This relation will sufficiently improve our understanding of those domains of physics where a stochastic web is mentioned. At the same time, we shall notice universality in the manifestations of quasi-symmetry in physical objects as different as the phase portrait of a dynamic system in phase space and the geometrical pattern of a steady-state flow of a liquid in coordinate space.

9.1 Stream lines in space

Stream lines of two-dimensional flows have a very simple structure and coincide with lines of the level of the stream function $\Psi(x, y)$. The behaviour of stream lines in steady-state three-dimensional flows can be completely different, since three-dimensional dynamics differs drastically from two-dimensional dynamics. The following equations

$$\frac{\mathrm{d}x}{v_x} = \frac{\mathrm{d}y}{v_y} = \frac{\mathrm{d}z}{v_z} \tag{9.1.1}$$

define stream lines of the field of velocities $v(x, y, z)$. A more convenient notation of the system (9.1.1), for example, in the following form

$$\frac{dx}{dz} = \frac{v_x}{v_z} \equiv f_1(x, y, z);$$

$$\frac{dy}{dz} = \frac{v_y}{v_z} \equiv f_2(x, y, z) \qquad (9.1.2)$$

shows that we are dealing with the 'non-steady-state' problem for a dynamic system with two-dimensional phase space (x, y). Variable z is playing the part of time. For fields with div $v = 0$, it is possible to present the system (9.1.2) in the Hamiltonian form in order to apply the already well-developed apparatus of the theory of dynamic systems to the full (Note 9.1).

If we take variable z in the system (9.1.2) to be 'time' (the part of time could be played by any other variable as well), the location of stream lines in space is determined by a family of two two-parametrical functions

$$x = x(z; x_0, y_0);$$

$$y = y(z; x_0, y_0). \qquad (9.1.3)$$

Here (x_0, y_0) are the 'initial' coordinates of a stream line, i.e., the values of (x, y) on a certain initial plane $z = z_0$. Therefore, expressions (9.1.3) satisfy the equations of motion (9.1.2) and the following initial condition

$$x_0 = x(z_0; x_0, y_0);$$

$$y_0 = y(z_0; x_0, y_0). \qquad (9.1.4)$$

Therefore, expressions (9.1.3) and (9.1.4) define a line in space (x, y, z) for a given field of velocities $v(x, y, z)$.

One or the other form of functions f_1 and f_2 in (9.1.2) (or, otherwise, of the stationary flow $v(x, y, z)$) may lead to stream lines with different spatial topology. Here, as well as in the case of trajectories of dynamic systems, we can speak of stable and unstable stream lines.

Let, for example, a stream line of the field of velocities v issue from a point r_1 and reach another point r_2 (Fig. 9.1.1). The points are connected by the following relation:

$$r_2 = \hat{\mathscr{L}} r_1 \qquad (9.1.5)$$

where operator $\hat{\mathscr{L}}$ defines the motion of a stream line. The case when, for small perturbations of δr_1, δr_2 is also small (Fig. 9.1.1a) corresponds to stable stream lines. Since the number of stream lines within a stream tube is conserved, the emerging chaos means that they get extremely entangled in space (Fig. 9.1.1*b*), as compared to the regular behaviour (Fig. 9.1.1*a*). Stream lines get intermingled in coordinate space in the same way as particle trajectories in the phase space of a Hamiltonian system.

If the flow is two-dimensional, there is no more *z*-dependency in (9.1.2) and, consequently, no chaos of stream lines. Certain types of steady-state three-dimensional flows also rule out the existence of chaos [5]. Further we shall mostly discuss flows which satisfy the so-called Beltrami condition

$$v = c \operatorname{rot} v \qquad (9.1.6)$$

where c is a function of coordinates. We shall also restrict ourselves to the case of $c = \pm 1$.

The meaning of condition (9.1.6) becomes clear from the stationary equation (8.1.3) for velocity

$$v \times \operatorname{rot} v = \nabla \Phi \qquad (9.1.7)$$

where $\Phi = v^2/2 + P/\rho$ is an arbitrary scalar function of coordinates and the full mechanical energy of liquid per unit mass. If Φ varies in space, i.e., the following equation

$$\Phi(x, y, z) = \text{const}$$

defines a certain surface, then, by means of a scalar multiplication of (9.1.7) by velocity v we get

$$(v\nabla)\Phi = 0.$$

Fig. 9.1.1 Stable (*a*) and unstable (*b*) stream lines.

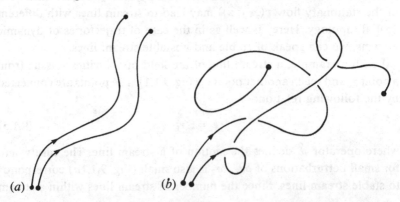

(*a*)   (*b*)

This condition means that function Φ is the first integral of the set of equations (9.1.1) for stream lines, i.e., that such flows rule out the existence of chaos. On the contrary, if in a certain region

$$\nabla\Phi = 0$$

then from (9.1.7) it follows that the field of velocities v satisfies the Beltrami condition (9.1.6) and, as will become clear later, chaos of stream lines is possible in such flows.

Now let us turn our attention to the equations of trajectories of liquid particles

$$\frac{dr}{dt} = v(r). \tag{9.1.8}$$

Since, in the case of a stationary field of velocities, stream lines coincide with trajectories of particles, chaos of stream lines implies also the stochastic behaviour of passive particles. Such behaviour of trajectories is called the Lagrangian turbulence.

Let $r(t)$ be the trajectory corresponding to the initial condition $r(t = 0) = r_0$. Let us find the equation of perturbation of this trajectory. With this aim in view, consider a close trajectory $r(t) + \delta r(t)$ with the initial condition $r(t = 0) = r_0 + \delta r_0$ where δr_0 is an infinitesimal vector. Then the perturbation of the trajectory $\delta r(t)$ will satisfy the following equation

$$\frac{d\delta r}{dt} = (\delta r \nabla) v(r)$$

which can be rewritten in the following form:

$$\frac{\partial}{\partial t} \delta r + (v\nabla)\delta r = (\delta r \nabla)v. \tag{9.1.9}$$

The stochasticity of stream lines and trajectories of particles of a liquid means that the Lyapunov exponent is positive

$$\lambda = \lim_{t \to \infty} \frac{1}{t} \ln \frac{|\delta r(t)|}{|\delta r_0|}. \tag{9.1.10}$$

In other words, it implies the exponential divergence of adjacent elements of the liquid. This can lead to a non-trivial effect related to the generation of a magnetic field in the case of motion of a conducting liquid – or to the effect of a hydromagnetic dynamo [6–7]. The evolution of magnetic

field B in the case of the flow of an incompressible liquid is described
by the following equation of induction

$$\frac{\partial B}{\partial t} = \text{rot}[\,v \times B\,] + \frac{1}{R_m}\nabla^2 B \qquad (9.1.11)$$

where $R_m = L_0 v_0 / \nu_m$ is the magnetic Reynolds number, L_0 is the characteristic scale, v_0 is the characteristic velocity, and ν_m is the magnetic
viscosity. Formally setting $R_m = \infty$ in (9.1.11), we get that B and δr satisfy
the same equation (9.19) as δr. Therefore, if the Lyapunov exponent
(9.1.10) is non-zero, this automatically implies the local exponential
growth of the magnetic field. Nevertheless, the fact that the Lyapunov
exponent is positive might still be insufficient for the fast dynamo. This
statement can be illustrated with the help of the following example.
Consider a non-stationary plane vortex motion of an incompressible
liquid. Under certain conditions, stochastic dynamics and diffusion of
passive particles occur in it. However, according to the Kauling theory
of the ban on dynamo [6] which is well known from magnetic hydrodynamics, in such two-dimensional systems the dynamo-effect is impossible
(Note 9.2).

9.2 Stream lines of the ABC-flow

In 1965 V. I. Arnold [11] suggested that the following steady-state
three-dimensional flow

$$v_x = A \sin z + C \cos y$$

$$v_y = B \sin x + A \cos z \qquad (9.2.1)$$

$$v_z = C \sin y + B \cos x$$

has a nontrivial topology of stream lines, since it satisfies the Beltrami
condition (9.1.6): rot $v = v$. The numerical analysis carried out in [12, 13]
confirmed the peculiarity of the flow (9.2.1). At the end of Sect. 9.1, we
have already noted that chaos of stream lines may lead to generation of
a magnetic field by a certain flow in a conducting liquid. This problem
was studied by Childress [14]. The flow (9.2.1) was named the ABC-flow
(Arnold–Beltrami–Childress) (Note 9.3).

Let us mention two prominent features of the ABC-flow. First, it is
the solution of the Navier–Stokes equation

$$\frac{\partial v}{\partial t} + (v\nabla)v = -\nabla(P/\rho) + \nu\nabla^2 v + F \qquad (9.2.2)$$

provided force F is chosen accordingly:

$$F = \nu v. \tag{9.2.3}$$

Equality (9.2.3) follows from the fact that the ABC-flow (9.2.1) satisfies the condition $\nabla^2 v = -v$. Besides, we know of the stability of the ABC-flow, in the case of a sufficiently large viscosity [19]. Therefore, the ABC-flow can occur as the result of an external forcing just as happened in the two-dimensional models of Sect. 8.2.

Another important characteristic of the ABC-flow is the fact that the set of equations

$$\frac{dx}{A \sin z + C \cos y} = \frac{dy}{B \sin x + A \cos z} = \frac{dz}{C \sin y + B \cos x}, \tag{9.2.4}$$

defining stream lines of the velocity field (9.2.1), can be presented in the explicit Hamiltonian form [20]. In order to do this, let us write (9.2.4) in the following form:

$$\frac{dx}{dz} = \frac{1}{\Psi} \frac{\partial}{\partial y} H; \qquad \frac{dy}{dz} = -\frac{1}{\Psi} \frac{\partial}{\partial x} H, \tag{9.2.5}$$

where

$$\begin{aligned} \Psi(x, y) &= C \sin y + B \cos x \\ H(x, y, z) &= \Psi(x, y) + A(y \sin z - x \cos z), \end{aligned} \tag{9.2.6}$$

and transfer from the independent variables x, y to new independent variables $\xi = \xi(x, y)$; $p = p(x, y)$:

$$\begin{aligned} \frac{dp}{dz} &= -\frac{1}{\Psi} \frac{\partial H}{\partial \xi} \left\{ \frac{\partial \xi}{\partial x} \frac{\partial p}{\partial y} - \frac{\partial \xi}{\partial y} \frac{\partial p}{\partial x} \right\}; \\ \frac{d\xi}{dz} &= \frac{1}{\Psi} \frac{\partial H}{\partial p} \left\{ \frac{\partial \xi}{\partial x} \frac{\partial p}{\partial y} - \frac{\partial \xi}{\partial y} \frac{\partial p}{\partial x} \right\}. \end{aligned} \tag{9.2.7}$$

By setting in (9.2.7)

$$\xi = x; \qquad p = \int_0^y dy' \, \Psi(x, y'), \tag{9.2.8}$$

we reduce the system to the following canonical form

$$\frac{dx}{dz} = \frac{\partial H}{\partial p}; \qquad \frac{dp}{dz} = -\frac{\partial H}{\partial x}, \tag{9.2.9}$$

where function

$$H(p, x, z) = C \sin y(p, x) + B \cos x + A[y(p, x) \sin z - x \cos z] \tag{9.2.10}$$

is obtained by substituting the expression for Ψ (9.2.6) into (9.2.8). This yields

$$p = By \cos x + C(1 - \cos y). \qquad (9.2.11)$$

The set of equations (9.2.9) with the Hamiltonian (9.2.10), generally speaking, is non-integrable, with the exception of an obvious case when it is reduced to the two-dimensional set (i.e., when one of the coefficients A, B, C becomes zero). To be more definite, assume that $A = 0$, we find the first integral in this – integrable – case:

$$C \sin y + B \cos x = H_0 = \text{const}. \qquad (9.2.12)$$

Stream lines on the plane (x, y) are shown in Fig. 9.2.1. There are three types of stream lines in it: closed stream lines, infinite stream lines and singular stream lines, passing through saddle points of the surface $H_0(x, y)$ and corresponding to separatrices. With the help of the integral (9.2.12), the system (9.2.9) can be easily integrated, its solution expressed in elliptic functions [13].

What happens to stream lines when $A \neq 0$? Let us consider the Hamiltonian of stream lines (9.2.10). If $A = 0$ (a two-dimensional case), it defines a family of cylindrical surfaces (stream tubes) corresponding to various values of the energy integral $H_0 = \text{const}$ (which is also the stream function). A perturbation of the Hamiltonian (9.2.10) in the case of small A is equivalent to a small non-stationary perturbation of the dynamic system. A considerable proportion of stream tubes slightly change their

Fig. 9.2.1   An integrable case of the ABC-flow at $A = 0$: stream lines on the plane (x, y).

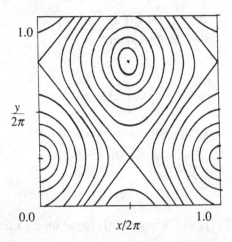

$\dfrac{y}{2\pi}$

0.0   1.0

$x/2\pi$

shapes in accordance with the Kolmogorov–Arnold–Moser theory. However, in the case of $A \neq 0$, there are such singular separatrix surfaces, which are heavily affected even by a small perturbation. The latter leads to formation of stochastic layers in the vicinity of destroyed separatrices and, consequently, to chaos of stream lines. In the case of large values of $A \sim 1$, stochastic layers expand (Fig. 9.2.2) and chaos of stream lines embraces a considerable portion of three-dimensional space. The appearance of large regions of chaos of stream lines in the ABC-flow, discovered in [12], is, in fact, the manifestation of a far more global phenomenon. We shall dwell upon this problem in Sect. 9.4.

9.3 Three-dimensional flows with symmetry and quasi-symmetry

As we have already noted in the previous chapter, quasi-crystal symmetry can, generally speaking, appear in two-dimensional hydrodynamics provided we create a stability region with the help of a source and viscous terms in such a flow.

Fig. 9.2.2 An example of the Poincaré cross-section of the ABC-flow stream lines at $z = 0$ for the case of $A = \sqrt{3}$, $B = \sqrt{2}$, $C = 1$ (Henon [12]).

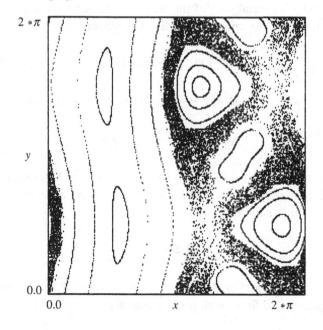

Three-dimensional steady-state flows also allow for the existence of quasi-symmetry [21, 22]. Let us consider the field of velocities defined by the following expressions:

$$v_x = -\frac{\partial \Psi}{\partial y} + \varepsilon \sin z$$

$$v_y = \frac{\partial \Psi}{\partial x} - \varepsilon \cos z \qquad (9.3.1)$$

$$v_z = \Psi$$

where ε is a parameter, while function $\Psi \equiv \Psi(x, y)$ satisfies the two-dimensional Helmholtz equation (8.1.11):

$$\nabla^2 \Psi + \Psi = 0. \qquad (9.3.2)$$

By direct testing, we can make sure that the flow (9.3.1) satisfies the incompressibility condition

$$\text{div } v = 0$$

and the Beltrami condition in the following form

$$\text{rot } v = -v. \qquad (9.3.3)$$

Let us consider a special case of the flow (9.3.1), where, as the solution of equation (9.3.2), we choose the following expression:

$$\Psi = \Psi_0 \sum_{j=1}^{q} \cos(\boldsymbol{R}\boldsymbol{e}_j) \qquad (9.3.4)$$

in which, as in (8.1.12), $\boldsymbol{R} = (x, y)$ is the coordinate vector and

$$e_j = \left\{ \cos \frac{2\pi}{q} j, \sin \frac{2\pi}{q} j \right\}$$

are unit vectors forming a regular q-star. Let us say, as before, that the flow (9.3.1) has the symmetry of the qth order, or simply quasi-symmetry.

Note that the function (8.1.12)

$$\Psi = \sum_{j=1}^{q} C_j \cos(\boldsymbol{R}\boldsymbol{e}_j + \alpha_j) \qquad (9.3.5)$$

as well as (9.3.4), satisfies the Helmholtz equation (9.3.2). It helps to describe generalized flows with quasi-symmetry.

The field of velocities, being a mirror image of (9.3.1), is defined by the following expressions

$$v_x = \frac{\partial \Psi}{\partial y} + A \sin z$$

$$v_y = -\frac{\partial \Psi}{\partial x} + A \cos z \qquad (9.3.6)$$

$$v_z = \Psi$$

where A is a parameter and function Ψ is the solution of equation (9.3.2). If we choose function Ψ in the form of (9.3.5), the flow possesses quasi-symmetry but its polarization sign is the reverse of (9.3.1), i.e., the Beltrami condition can be written in the following form

$$\operatorname{rot} v = v.$$

Specifically, if

$$\Psi = C \sin y + B \cos x,$$

the corresponding field of velocities is the ABC-flow. Thus, flows (9.3.1) and (9.3.6) are appropriate generalizations of the ABC-flow as applied to an arbitrary quasi-symmetry.

For the quasi-symmetrical flow (9.3.1), the set of equations (9.1.2) defining stream lines has the following form

$$\frac{dx}{dz} = -\frac{1}{\Psi} \frac{\partial}{\partial y} H(x, y, z); \qquad \frac{dy}{dz} = \frac{1}{\Psi} \frac{\partial}{\partial x} H(x, y, z), \qquad (9.3.7)$$

where the following expression

$$H(x, y, z) = \Psi(x, y) + \varepsilon V(x, y, z);$$

$$V(x, y, z) = -x \cos z - y \sin z \qquad (9.3.8)$$

is the Hamiltonian defining the equations of motion (9.3.7). In the next section we will show how dynamic systems (9.3.7) and (9.3.8) can be reduced to a typical canonical form. However, even now certain results of the theory of dynamic systems can be applied directly to equations (9.3.7).

The non-perturbed part of the system (9.3.7) with the Hamiltonian (9.3.8) at $\varepsilon = 0$ has numerous singular trajectories passing through singular saddle points, or separatrices. For example, for $q = 4$ and $q = 3, 6,$

when function Ψ is defined by equations (8.1.16) and (8.1.17), all separatrices belong to one and the same value of the stream function Ψ, therefore constituting a single square or hexagonal network. Any arbitrarily small periodic perturbation along z (i.e., arbitrarily small ε) leads to the destruction of separatrices and to a finite region of stochastic dynamics of stream lines emerging in the vicinity of the destroyed separatrices. In our case, this means the appearance of a stochastic web with square or hexagonal cells in space (x, y, z). Examples of these webs are shown in Figs. 9.3.1 and 9.3.2. Small regions, or islands, of stable dynamics of stream lines, are still left in the web. This means that stream lines are wound regularly on a certain invariant surface (a stream tube). In our case, these are periodically bending tubes with the axis pointing in several directions which depend on the order of symmetry q. Cross-sections of these tubes form windows in the stochastic web in Fig. 9.3.3. The remaining portion of the web is filled by a single trajectory and is a fractal manifold occupying a finite volume in space (x, y, z).

As it nears the points of stagnation, or saddles, lying on separatrix intersections, a stream line, following a random law, approaches the direction of its furthest motion. As a result, a spatial diffusion process occurs, analogous to the Brownian motion of a particle along a square or hexagonal grid. Two examples of such diffusion, for $q = 4$ and $q = 3$, obtained in the course of a numerical analysis, are presented in Fig.

Fig. 9.3.1 Stream lines form a stochastic web with square symmetry ($q = 4$). The results of numerical calculations of equations (9.3.1) for $q = 4$ and $\varepsilon = 0.6$ are presented here. The plane of Poincaré cross-section (x, y) corresponds to (*a*) $z = 0$; (*b*) $z = \pi/4$. The size of the square is $8\pi \times 8\pi$. Reprinted by permission from *Nature*, **337**, 113 [22]. Copyright © 1989 Macmillan Magazines Ltd.

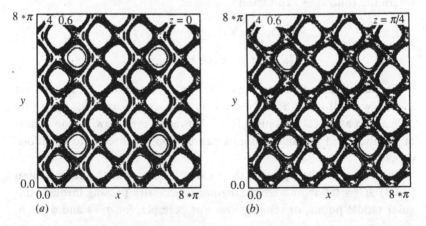

(*a*)   (*b*)

9.3.4. A stream line may cover a considerable ground on the plane $z = $ const, then turn and go over to the plane $x = $ const tracing a rather complicated loop, and finally return to the initial plane $z = $ const. Therefore, looking at the picture of diffusion we observe jumps of stream lines on the plane $z = $ const. This is seen very clearly in the case of $q = 4$ (Fig. 9.3.4a).

In essence, the results obtained in this section prove that the structure of a steady-state three-dimensional flow is rather non-trivial, although the formal expression for the field of velocities may look quite simple. The regions of stochasticity of stream lines divide the whole coordinate space in which the flow takes place into a network of cells which can have a simple crystal symmetry ($q = 3, 4, 6$) as well as a very complex, quasi-crystal one (for example, $q = 5, 7, \ldots$). The thickness of stochasticity regions, or of a stochastic web, is determined by the parameters of the flow). It is natural to call this phenomenon the turbulence of stream lines.

In this section we have shown much more than simply the process of formation of the spatial stochastic web in hydrodynamic steady-state flows. A hydrodynamic web determines the pattern of the flow. It is most likely that many other flows with fairly complex patterns, besides the flow (9.3.1), generate the Lagrangian turbulence. Therefore, the bifurcation of pattern also implies bifurcational changes of the hydrodynamic web.

Fig. 9.3.2   A hexagonal stochastic web. The results of numerical computations of equation (9.3.1) at $q = 3$. The plane of Poincaré cross-section (x, y) corresponds to (a) $\varepsilon = 0.3$, $z = \pi/4$; (b) $\varepsilon = 1.0$, $z = 0$. The size of a square is $9.2\pi \times 9.2\pi$. Reprinted by permission from *Nature*, **337**, 113 [22]. Copyright © 1989 Macmillan Magazines Ltd.

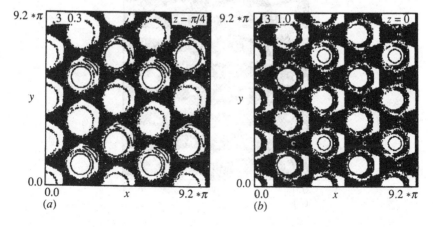

In the next section we discuss the theory of formation of the hydrodynamic stochastic web.

9.4 Stochastic layers and stochastic webs in hydrodynamics

Let us consider the theory of formation of a hydrodynamic stochastic
web for small values of $\varepsilon \ll 1$. We shall give its full presentation for the
case of $q = 3$. The same procedure can be applied to other values of q.

Fig. 9.3.3 Stratified structure of a hexagonal web $(q = 3)$. The
Poincaré cross-section in the vertical plane (y, z) corresponds to (a)
$x = 0$, $\varepsilon = 1$; (b) $x = \pi$, $\varepsilon = 1$. The size of the square is $8\pi \times 8\pi$.
Reprinted by permission from *Nature*, **337**, 113 [22]. Copyright ©
1989 Macmillan Magazines Ltd.

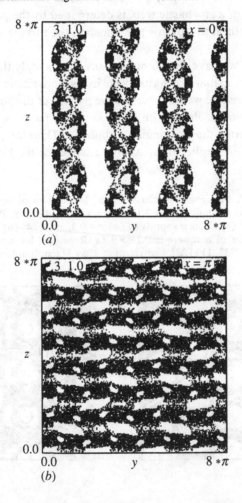

Now, using a non-linear transform of the type of (9.2.8), we transfer from the independent variables (x, y) in equations (9.3.7) to new canonical variables (x, p):

$$p \equiv p(x, y) = \int_0^y \mathrm{d}y'\, \Psi(x, y'). \qquad (9.4.1)$$

Expressed in the variables (x, p), equations (9.3.7) can be written in the

Fig. 9.3.4   The general sketch of space diffusion of a stream line in the plane (x, y) for $z = 0$; (a) square symmetry $q = 4$, $\varepsilon = 0.6$, $-400\pi < x < 200\pi$, $-200\pi < y < 400\pi$; (b) hexagonal symmetry $q = 3$, $\varepsilon = 0.003$, $-30\pi < x < 30\pi$, $-30\pi < y < 30\pi$. Reprinted by permission from *Nature*, **337**, 113 [22]. Copyright © 1989 Macmillan Magazines Ltd.

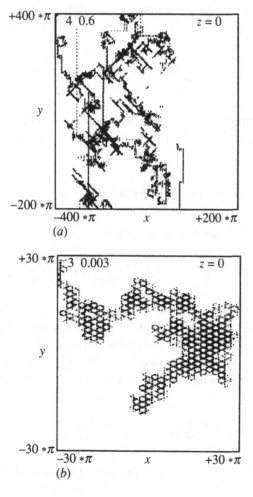

canonical form

$$\frac{dx}{dz} = -\frac{\partial H}{\partial p}; \qquad \frac{dp}{dz} = \frac{\partial H}{\partial x} \qquad (9.4.2)$$

with the Hamiltonian

$$H(p, x, z) = \Psi(x, y(p, x)) - \varepsilon y(p, x)\sin z - \varepsilon x \cos z, \qquad (9.4.3)$$

which is a periodic function of 'time' z.

Thus, by means of the transform (9.4.1) the problem of determining stream lines in three-dimensional steady-state flows is reduced to the analysis of the Hamiltonian system with one and a half degrees of freedom.

Let us present an explicit form of the nonlinear transform (9.4.1) in the case of hexagonal symmetry ($q = 3$) when the non-perturbed Hamiltonian is defined by equation (8.1.17):

$$\Psi = \cos x + \cos\left(\frac{x}{2} + \frac{\sqrt{3}}{2}y\right) + \cos\left(\frac{x}{2} - \frac{\sqrt{3}}{2}y\right). \qquad (9.4.4)$$

To be more definite, we have set the coefficient $\Psi_0 = 1$. In this expression, variable y as a function of x and p is found by solving the non-linear equation (9.4.1):

$$p = y\cos x + \frac{4\sqrt{3}}{3}\cos\frac{x}{2}\sin\frac{\sqrt{3}}{2}y. \qquad (9.4.5)$$

Solving (9.4.5) with respect to y, we obtain

$$y(x, p) = \frac{p}{\cos x} + \frac{4\sqrt{3}}{3}\sum_{n=1}^{\infty}\frac{(-1)^n}{n}J_n\left(\frac{n\cos x/2}{2\cos x}\right)\sin\left(\frac{\sqrt{3}\,np}{2\cos x}\right). \qquad (9.4.6)$$

By means of (9.4.6) the Hamiltonian of the problem (9.4.3) in the case of $q = 3$ can be written in an explicit form as a function of x, p and z.

Now let us apply the method of canonical transforms. With the help of the generating function, we transfer from the variables (x, p) to canonical variables (J, φ) which are action-angle variables for the non-perturbed problem ($\varepsilon = 0$):

$$J = \frac{1}{2\pi}\oint dx\,p; \qquad \varphi = \frac{\partial S(J, x)}{\partial J}. \qquad (9.4.7)$$

Expressing p through x and Ψ with the help of (9.4.4) and (9.4.5) and substituting it into (9.4.7) we get

$$S(x, J) = \frac{2\sqrt{3}}{3} \int_0^x dx \left\{ \cos x \cdot \arccos\left(\frac{\Psi - \cos x}{2 \cos x/2}\right) \right.$$

$$\left. + 2 \cos \frac{x}{2} \left[1 - \left(\frac{\Psi - \cos x}{2 \cos x/2}\right)^2 \right]^{1/2} \right\}. \tag{9.4.8}$$

There are two intervals with different expressions for the action and the angle variable. At $-1 < \Psi \leq 3$, it follows from (9.4.8) that

$$J = \frac{4\sqrt{3}}{3\pi} \int_0^\Psi dh \frac{h}{(2h+3)^{1/4}} K[\varkappa_1(h)] \tag{9.4.9}$$

where

$$\varkappa_1(h) = \frac{[3 - h^2 + 2(2h+3)^{1/2}]^{1/2}}{2(2h+3)^{1/4}}$$

is the modulus of the full elliptic integral $K(\varkappa_1)$. From equation (9.4.9) there follows the expression for the frequency of non-linear oscillations

$$\Omega = \frac{\partial \Psi}{\partial J} = \frac{\pi\sqrt{3}}{4\Psi} (2\Psi + 3)^{1/4} K^{-1}[\varkappa_1(\Psi)]. \tag{9.4.10}$$

In the vicinity of a separatrix ($\Psi \to -1$), the expression for $\Omega(\Psi)$, as usual, has a logarithmic singularity

$$\Omega = \frac{\pi\sqrt{3}}{6} \ln^{-1}(1/|\Psi + 1|). \tag{9.4.11}$$

Within the region of values $-\frac{3}{2} \leq \Psi < 1$ instead of (9.4.9) we have

$$J = \frac{8\sqrt{3}}{3\pi} \int_0^\Psi dh\, h\{[1 + (2h+3)^{1/2}]^2 - (h+1)^2\}^{-1/2} K[\varkappa_2(h)] \tag{9.4.12}$$

where $\varkappa_2 = 1/\varkappa_1$. The frequency of non-linear oscillations near a separatrix in this region of values of the non-perturbed Hamiltonian Ψ is, as before, defined by equation (9.4.11).

The Hamiltonian of stream lines (9.4.3) at $\varepsilon = 0$ (the two-dimensional case) defines the family of cylindrical surfaces corresponding to various values of the energy integral $\Psi = \text{const} \equiv E$. They include singular separatrix surfaces which are strongly affected even by small perturbations. Their cross-section by the plane $z = \text{const}$ in the case of $q = 3$ yields a hexagonal net which is destroyed at $\varepsilon \neq 0$, a stochastic web forming in its place.

Let us determine the thickness of a stochastic web. With this aim in view, consider the Hamiltonian (9.4.3) at $\varepsilon \ll 1$. The change of the Hamiltonian Ψ can be found with the help of the following equation

$$\frac{d\Psi}{dz} = \frac{\varepsilon}{\Psi} \left(\frac{\partial \Psi}{\partial x} \sin z - \frac{\partial \Psi}{\partial y} \cos z \right). \tag{9.4.13}$$

Separatrix surfaces at $q = 3$ correspond to the value of the Hamiltonian $\Psi = -1$ and are defined by the following equations

$$x = \pi(2n_1 + 1); \qquad x = \sqrt{3}\, y + 2\pi(2n_2 + 1);$$
$$x = -\sqrt{3}\, y + 2\pi(2n_3 + 1); \qquad (n_1, n_2, n_3 = 0, \pm 1, \ldots). \tag{9.4.14}$$

Let us find one of the separatrix solutions of the non-perturbed problem $(\varepsilon = 0)$. If, for example, $x = \pi$, then from (9.3.7) we obtain the following

$$\frac{dy}{dz} = \cos \frac{\sqrt{3}}{2}\, y. \tag{9.4.15}$$

Integration of (9.4.15) yields

$$\frac{\sqrt{3}}{2}\, y = \frac{\pi}{2} - 2 \arctan \exp\left[-\frac{\sqrt{3}}{2}(z - z_n) \right], \tag{9.4.16}$$

where z_n is the instant of piercing the surface $y = 0$. Expression (9.4.16) has a simple meaning. It is a soliton of the stream line forming a hexagonal pattern. Substituting (9.4.16) into (9.4.13) and integrating over z we find the change of the non-perturbed energy $\Psi = E$ in the course of motion between two adjacent saddle points $z^{(1)}$ and $z^{(2)}$:

$$\Delta E = \varepsilon \int_{z^{(1)}}^{z^{(2)}} dz \frac{\sin z}{\cosh \sqrt{3}(z - z_n)/2} \approx \int_{-\infty}^{\infty} dz \frac{\sin z}{\cosh \sqrt{3}(z - z_n)/2}. \tag{9.4.17}$$

Hence it follows that

$$\Delta E \approx \sqrt{3}\, \pi \varepsilon \sin z_n. \tag{9.4.18}$$

Equations (9.4.18) and (9.4.11) give us the mapping describing the behaviour of stream lines in the vicinity of a separatrix

$$E_{n+1} = E_n + \sqrt{3}\, \pi \varepsilon \sin z_n,$$
$$z_{n+1} = z_n + \frac{2\sqrt{3}}{3} \ln(1/|1 + E_{n+1}|). \tag{9.4.19}$$

For a region within the stochastic layer in (9.4.19) we have (see Sect. 3.2)

$$|1 + E| \leqslant 2\pi\varepsilon. \tag{9.4.20}$$

Hence, we find the thickness of the stochastic layer $2\delta E$, provided we set $E = -1 \pm \delta E$ in (9.4.20) and choose the equality sign

$$2\delta E = 4\pi\varepsilon. \qquad (9.4.21)$$

The above result shows that in space (x, y, z) there is a finite region – the web of thickness of the order of ε where stream lines are stochastically arranged in space. Although we came to this conclusion in the specific case of hexagonal symmetry, it can fairly easily be generalized to the case of the ABC-flow [21] or a flow with an arbitrary symmetry. This means that, when forming a certain pattern in space, steady-state flows, at the same time, produce regions of stochastic arrangement of stream lines with thickness of the order of ε in the vicinity of this pattern. Three-dimensionality of patterns is a necessary condition for the formation of structured chaos.

In Chapter 6 we described the mechanism of the formation of a web with quasi-symmetry. But here we have met a completely different form of such a web. A quasi-symmetrical hydrodynamic web forms not in the phase space of a particle, but in the usual coordinate space (x, y, z) where

Fig. 9.4.1   A stochastic web of stream lines with fifth-order quasi-symmetry in the plane $z = 0$. The solution of equations (9.3.1) corresponds to $q = 5$ and $\varepsilon = 0.03$. The size of the square is $24\pi \times 24\pi$. Reprinted by permission from *Nature*, **337**, 113 [22]. Copyright © 1989 Macmillan Magazines Ltd.

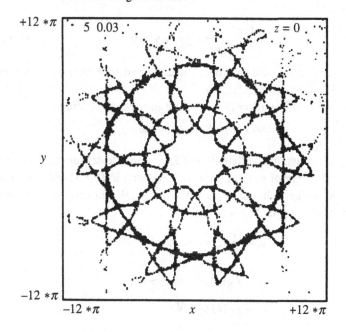

the part of time is played by the variable z. This web is three-dimensional and has quasi-symmetry in the plane (x, y). An example of a three-dimensional flow, in which a plane pattern with fifth-order quasi-symmetry appears, is shown in Fig. 9.4.1. It was obtained numerically for the system (9.3.1) with the stream function (9.3.4) at $\Psi_0 = 1$ and $q = 5$.

9.5 Helical steady-state flows

In this section we are going to introduce another type of three-dimensional steady-state flow with helical symmetry of the order of N which is a three-dimensional generalization of a plain flow with the stream function Ψ satisfying the Helmholtz equation (9.3.2). These flows are symmetrical with regard to rotation around the z-axis by the angle $2\pi/N$. They owe their existence to a presentation of the field of velocity (9.3.1) in cylindrical coordinates

$$
\begin{aligned}
v_r &= -\frac{1}{r}\frac{\partial\Psi}{\partial\varphi} + \frac{\varepsilon}{r}\sin z \\[6pt]
v_\varphi &= \frac{\partial\Psi}{\partial r} - \frac{\varepsilon}{r}\cos z \\[6pt]
v_z &= \Psi.
\end{aligned}
\qquad (9.5.1)
$$

Here the function $\Psi \equiv \Psi(r, \varphi)$ satisfies equation (9.3.2) written in polar coordinates:

$$
\frac{1}{r}\frac{\partial}{\partial r}r\frac{\partial\Psi}{\partial r} + \frac{\partial^2\Psi}{\partial\varphi^2} + \Psi = 0.
\qquad (9.5.2)
$$

We can easily make sure that the Beltrami condition $\operatorname{rot} v = -v$ is valid for the field of velocity (9.5.1).

The solution of equation (9.5.2) is the following function

$$
\Psi(r, \varphi) = \sum_n C_n J_n(r) \cos n\varphi,
\qquad (9.5.3)
$$

where C_n are constant coefficients and $J_n(r)$ is the Bessel function. Several cases of (9.5.3) are of special interest. Retaining in (9.5.3) only the term with $n = N$, we get a helical flow with rotational symmetry in the plane $z = \text{const}$ by the angle $2\pi/N$:

$$
\begin{aligned}
v_r &= \frac{N}{r}J_N(r)\sin N\varphi + \frac{\varepsilon}{r}\sin z \\[6pt]
v_\varphi &= J'_N(r)\cos N\varphi - \frac{\varepsilon}{r}\cos z \\[6pt]
v_z &= J_N(r)\cos N\varphi.
\end{aligned}
\qquad (9.5.4)
$$

The set of equations for stream lines can be written in the following form

$$\frac{dr}{dz} = \frac{1}{\Psi}\left\{-\frac{1}{r}\frac{\partial\Psi}{\partial\varphi} + \frac{\varepsilon}{r}\sin z\right\}$$

$$r\frac{d\varphi}{dz} = \frac{1}{\Psi}\left\{\frac{\partial\Psi}{\partial r} - \frac{\varepsilon}{r}\cos z\right\},$$

(9.5.5)

where

$$\Psi(r, \varphi) = J_N(r)\cos N\varphi$$

(9.5.6)

is the first integral if the perturbation at $\varepsilon = 0$ is absent. Equations (9.5.5) can also be presented in the Hamiltonian form with the following Hamiltonian

$$H(r, \varphi, z) = \Psi(r, \varphi) - \varepsilon(\varphi \sin z + \ln r \cos z).$$

(9.5.7)

Separatrices of a non-perturbed system with the Hamiltonian (9.5.6) constitute a regular web (Fig. 9.5.1). In a pair of cells symmetrical with regard to the centre, the liquid is rotating in the opposite directions. We find the web in the helical flows (9.5.4) just as in the flows with quasi-symmetry. A perturbation in the case of $\varepsilon \neq 0$ destroys a separatrix network and creates a stochastic layer in its place.

Fig. 9.5.1 The lines of the level of the stream function (9.5.6) for a helical field with fifth-order symmetry.

A special case of the helical flow (9.5.4) is a cylindrical flow at $N = 0$. In this case, from (9.5.6) it follows that

$$\Psi = J_0(r) \qquad\qquad (9.5.8)$$

and the problem is integrable. Indeed, from (9.5.5) it follows that

$$\frac{dr}{dz} = \frac{\varepsilon \sin z}{r J_0(r)} \qquad\qquad (9.5.9)$$

and this equation can be immediately integrated

$$\varepsilon \cos z + r J_1(r) = \text{const.} \qquad\qquad (9.5.10)$$

Therefore, the rotational moment of these flows is conserved:

$$v_\varphi r = -\text{const},$$

while they describe the motions of a liquid with the singularity on the axis $r = 0$, resembling the vortex motion in tornadoes.

We can draw an analogy between the patterns formed by hydrodynamic flows with helical symmetry and the patterns on the phase plane described in Sect. 5.2.

9.6 The stochasticity of stream lines in a stationary Rayleigh–Benard convection

The above situation seems to be fairly universal, so that many well-known patterns in other steady-state motions in liquid, gas or plasma must have similar properties. This means that patterns emerging in a liquid prior to the appearance of turbulence, form a stochastic web of stream lines. Besides, the arrangement of this web in space reflects the type of pattern [23]. This conclusion is based on the following. Let us turn to any suitable method of analysis of hydrodynamic matter in a pre-turbulent state. This state can be described by a finite number of harmonics obtained as the result of expansion of the solution into a Fourier series with their subsequent cutting off. Any finite series of harmonics possesses the property of quasi-symmetry. If the problem is three-dimensional, it corresponds to a dynamic system with three-dimensional phase space. Therefore, the stochasticity of stream lines must occur in almost any general situation of three-dimensional flows characterized by quasi-periodicity.

Let us illustrate the above with the example of stream lines in the case of a stationary Rayleigh–Benard convection in a layer of liquid which is infinite in the horizontal direction. If the Rayleigh number R, which is a dimensionless characteristic of the difference between the temperatures on the layer boundaries, excludes the critical number R_c, the liquid forms hexagonal Benard cells. For a small supercriticality $\varepsilon = (R - R_c)^{1/2} \ll 1$, the spatial dependency of the field of velocities v, with an accuracy up to the terms of the second order of smallness in parameter ε, has the following form [24]

$$v = \varepsilon \text{ rot rot } z\Psi, \qquad (9.6.1)$$

where

$$\Psi = \left[\cos x + \cos\left(\frac{x}{2} + \frac{\sqrt{3}}{2}y\right) + \cos\left(\frac{x}{2} - \frac{\sqrt{3}}{2}y\right)\right](\sin z + a\varepsilon \sin 2z)$$

$$+ b\varepsilon\left[\cos \sqrt{3}\, y + \cos\left(\frac{\sqrt{3}}{2}y + 3x\right) + \cos\left(\frac{\sqrt{3}}{2}y - 3x\right)\right]\sin 2z. \quad (9.6.2)$$

Explicit expressions for the coefficients a and b, depending upon the parameters of the problem were obtained in [24], while [25] provides us with an expansion with an accuracy up to third-order terms in ε.

Equations (9.6.2), defining the dynamics of a passive ingredient in the field of velocities (9.6.1), have the following form

$$\frac{dx}{dt} = -\left(\sin x + \sin\frac{x}{2}\cos\frac{\sqrt{3}}{2}y\right)(\cos z + 2\varepsilon_1 \cos 2z)$$

$$- 2\varepsilon_2 \cos\frac{\sqrt{3}}{2}y \sin\frac{3}{2}x \cos 2z$$

$$\frac{dy}{dt} = -\sqrt{3}\cos\frac{x}{2}\sin\frac{\sqrt{3}}{2}y(\cos z + 2\varepsilon_1 \cos 2z)$$

$$\qquad\qquad\qquad\qquad\qquad\qquad\qquad\qquad\qquad (9.6.3)$$

$$- \frac{2\sqrt{3}}{3}\varepsilon_2\left(\sin\sqrt{3}\,y + \sin\frac{\sqrt{3}}{2}y\cos\frac{3}{2}x\right)\cos 2z$$

$$\frac{dz}{dt} = \left(\cos x + 2\cos\frac{x}{2}\cos\frac{\sqrt{3}}{2}y\right)(\sin z + \varepsilon_1 \sin 2z)$$

$$+ \varepsilon_2\left(\cos\sqrt{3}\,y + 2\cos\frac{\sqrt{3}}{2}y\cos\frac{3}{2}x\right)\sin 2z,$$

where

$$\varepsilon_1 = a\varepsilon, \qquad \varepsilon_2 = 3b\varepsilon.$$

The picture of stream lines in hexagonal cells with the first-order approximation in parameter ε, i.e., when $\varepsilon_1 = 0$ and $\varepsilon_2 = 0$, was obtained in [26]. Stream lines are closed curves lying on vertical cylindrical surfaces (see Fig. 9.6.1).

With the second-order approximation in parameter ε, the picture is different. Stream lines are mostly non-closed and are wound around toroidal surfaces. Figure 9.6.2*a* presents the results of numerical integration of the system (9.6.3) in the cases of $\varepsilon_1 = 0$ and $\varepsilon_2 = 0.2$. The cross-section of one of the tori by the plane $z = \pi/2$ is presented. Near a cell's border the toroidal surfaces are destroyed. Figure 9.6.2*b* illustrates the chaos of stream lines in one of the sectors of a hexagonal cell. Let us note that with this approximation in parameter ε, despite the chaos emerging within the cells, a scalar ingredient cannot diffuse between the cells since on a cell's border the perpendicular component of velocity is zero [26]. In our case, the formation of a stochastic web is possible only if we take into account the terms of a higher order of smallness in ε.

Fig. 9.6.1   Stream lines in a hexagonal cell in the first approximation on the over-criticality parameter ε.

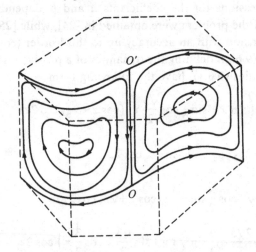

Fig. 9.6.2   The Poincaré cross-section of the system (9.6.3) at $z = \pi/2$; (*a*) a cross-section of toroidal surfaces; (*b*) a stochastic stream line in a cell.

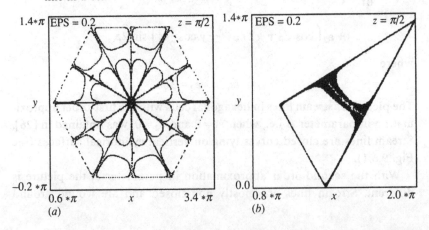

PART IV · Miscellanea

10
Patterns in art and nature

So far, we have been discussing various kinds of pattern with regular or almost regular symmetry. They emerged either in phase space of dynamic systems or in coordinate space of hydrodynamic flows. Common to all these cases was the method of obtaining or revealing patterns. Such patterns emerged not as the result of some artificial formal algorithm but as an expression of natural laws. In ancient times, however, people did not possess the level of knowledge available to us today. Perhaps it was the attempt to penetrate into the laws of creation of regular patterns, that gave rise to the art of ornament. Or perhaps this form of human activity had nothing to do with what was observed in nature. In either event, it would be interesting to make a number of comparisons between ancient ornaments and the pictures drawn by the trajectory of a real particle under certain conditions.

10.1 Two-dimensional tilings in art

Byzantine mosaic is one of the oldest examples of symmetrical periodic tilings of a plane (Fig. 10.1.1). Although the periodicity condition might have arisen as an independent problem, practical aims of architectural design required exactly this kind of ornament. Tiles of one shape (or of several different shapes) were to form the elementary components of a tiling. The element of an ornament was to be reproduced as many times as need, so that eventually any chosen portion of the plane could be paved.

The ornamental technique reached its peak of development in Muslim art. Elementary cells of an ornament became far more complex (Figs. 10.1.2 to 10.1.4). The periodicity condition required special ingenuity from the artist. Even the introduction of pentagonal (Fig. 10.1.2a) and heptagonal (10.1.3) elements into an ornament did not lead to a digression from the periodical pattern of the tiling.

211

In Central Asia there was a variety of methods enabling the artist to construct a regular pentagon or a star [2]. One is presented in Fig. 10.1.5. But so far, we know of no examples of aperiodic mosaics of the quasi-crystal type from this area.

If we take an interest exclusively in the geometrical features of ornaments, we should turn to the unique collection of various types of

Fig. 10.1.1   Examples of Byzantine mosaic (Wade [1]).

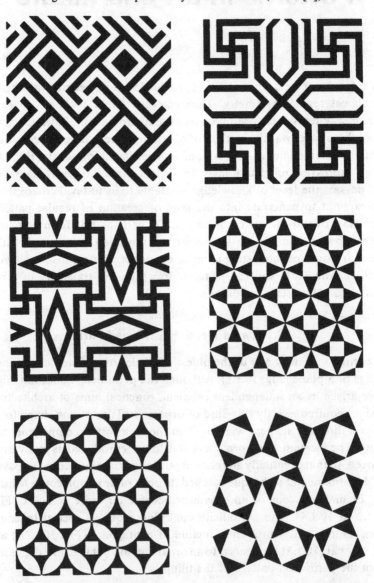

Fig. 10.1.2 Two tiles from Central Asia (12th century) and the techniques of their design (Bulatov [2]): (*a*) from the museum in Samarkand, (*b*) from the museum in Bukhara.

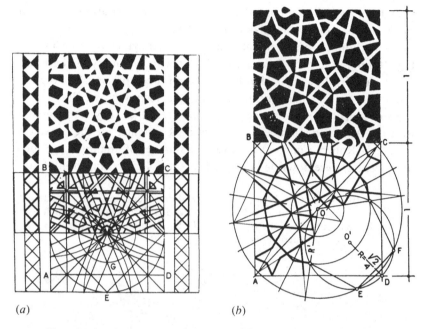

(*a*)   (*b*)

Fig. 10.1.3 An ornament from the Hodja Ahmad mausoleum and the technique of its design (Samarkand, Shah-i-Zinda, 14th century) (Bulatov [2]).

ornaments found in the Alhambra Palace in Granada, Spain. It was there that an ornament with fifth-order symmetry was discovered. Why did it happen there that an artist had the courage – or inspiration – to break away from periodic symmetry? And how did the artist arrive at such an idea? This is a puzzle unsolved to this day.

Some examples of ornaments with fifth-order symmetry are shown in Fig. 10.1.6*a*, *b*, and *c*. In *d*, an ornament with symmetry of the fourth order made by the same technique is shown for comparison. The pictures show that artists had freely mastered the technique of drawing an aperiodic pentagonal ornament. It is possible that they were familiar with the Ammann's lattice which served as a stencil for the ornament.

In 1977 Robert Ammann discovered a number of unknown aperiodic tilings which were described only in reference [4]. The Ammann's lattice

Fig. 10.1.4   One of the most widespread ornaments in Central Asia (the mausoleum of Shah-i-Zinda, Samarkand) (Hrbas and Knobloch [3]): (*a*) one cell of the ornament; (*b*) the formation of a periodic tiling based on such cells.

(*a*)

is formed by five sets of parallel lines tilted over angles $2\pi/5, 4\pi/5, \ldots$. Spacings between the lines correspond to Fibonacci numbers. This set of lines is clearly seen in Fig. 10.1.6c (see Sect. 7.1).

In comparison, the pattern on Central Asian ornaments was created with the help of the geometrical construction of a pentagon (Fig. 10.1.7). The basis for the mosaic shown is formed by only three elements (not counting the shapes of intermediate spaces between them). Here, an ornament was made not with the help of identical tiles with the same pattern, but by way of dense packing of several elements of a certain shape. The greatest achievement of the artists of the Alhambra was the practical demonstration of the existence of dense packing – in several

Fig. 10.1.4—continued

(b)

different ways. Thus, the artist's imagination gained much greater freedom compared to that he had enjoyed while working on periodic ornaments. Reference [5] draws an analogy between the process of creating a pentagonal tiling and a musical improvization on a given theme. This analogy could be extended much further if we allowed for the existence of defects in the tiling.

It is worth noting that Albrecht Dürer devised a unique method of constructing a regular pentagon. We also know of his attempts to pave a plane with pentagons and several other elements. However, he never discovered patterns of the Ammann's lattice type.

Results obtained in Chapters 6 and 7 provide the most widespread solution to problems concerning patterns with quasi-symmetry by reducing it to an algorithm. This solution lies in making a stencil with the help of the Hamiltonian (6.4.7). Let us demonstrate this by means of a simple example [6]. Figure 10.1.8 shows a typical decagonal element of a Muslim

Fig. 10.1.5 Construction of a pentagon with the help of a constant span of dividers, equal to the given length of the pentagon's diagonal (the Abu Bakr al-Haliq at-Takjir method) (Bulatov [2]).

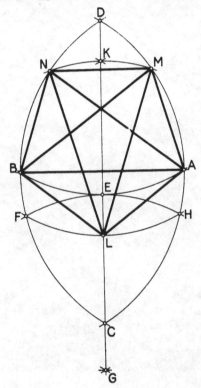

Fig. 10.1.6 Examples of non-periodic ornaments with fifth-order symmetry (*a, b, c*) and a periodic ornament with fourth-order symmetry (*d*) [1].

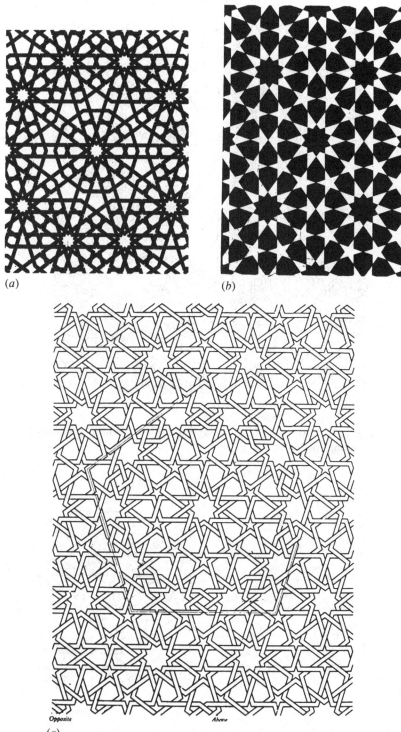

(*a*)

(*b*)

(*c*)

Opposite   *Above*

Fig. 10.1.6—continued

(d)

Fig. 10.1.7   An example of pentagonal tiling (a) and its geometric
decoding (b) (Critchlow [5]).

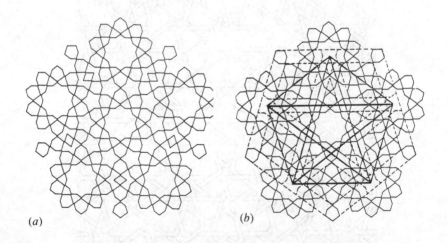

(a)   (b)

ornament and its 'decoding' with the help of a fifth-order symmetry relief which is already familiar to us. Superimposing the picture and the relief, we find the lines which led to the loss of pentagonal symmetry and to degeneracy of the ornament into a usual periodic pattern. Figure 10.1.9 demonstrates one possible way of obtaining an ornament with fifth-order symmetry using the same relief and preserving the same decagon as the main detail of the pattern.

A similar idea of the relief's utilization was discussed in Sect. 7.1 (see Fig. 7.1.3) in connection with the Penrose tiling. Boundless possibilities for new works of ornamental graphics are opened by applying the reliefs

Fig. 10.1.8 An example of a typical decagonal element of a Moslim mosaic in Tbilisi (*a*) and (*b*) its decoding with the help of a fifth-order symmetry relief (Zaslavsky *et al.* [6]).

(*a*)

defined by the following Hamiltonians:

$$H_q = \sum_{j=1}^{q} \cos(\boldsymbol{re}_j),$$

where $\boldsymbol{r}$ is a two-dimensional vector, while $\boldsymbol{e}_j$ are unit symmetrically arranged vectors. An even more tempting possibility is to use an ornament drawn by the trajectory of a particle performing a random walk within a stochastic web.

The connection between the art of ornament and the mathematical problem of tilings has a rich history and is well covered by specialist literature (see, for example, [4, 7, 8]). However slightly this subject is touched upon in our book, we could hardly ignore the art of Maurits Escher [9]. In many aspects his works attract the attention of mathematicians, art historians, designers and architects. This is exemplified, among other things, by the conference devoted to his art [10]. In many of his drawings, Escher also used one or several tiles to create

Fig. 10.1.8—continued

(b)

extremely complex ornaments belonging to the so-called colour symmetry (see, for example, Fig. 10.1.10 borrowed from [11]). By means of the same technique, Escher was trying to effect a transition from one symmetry to another. The best evidence of this is given by his Metamorphoses III (Fig. 10.1.11). There are certain places in it where the artist tried to 'sew together' the symmetries of the third and sixth order with that of the fourth order. However, nature can do a better job of sewing together different symmetries. This is demonstrated, for example, by Fig. 10.1.12. This figure is characteristic of a two-dimensional flow (Sect. 8.2) where the initial condition for the field of velocity had hexagonal symmetry, while the field of the external forcing possessed fifth-order symmetry.

Fig. 10.1.9   The use of relief with fifth-order symmetry to obtain an ornament with the same symmetry including the decagon in Fig. 10.1.8*a*.

Fig. 10.1.10 The use of two tiles to design a periodic ornament shown in the sketch, *Symmetry Drawing E126*, by Maurits Escher [11] © 1990 M. C. Escher Heirs/Cordon Art – Baarn – Holland.

Fig. 10.1.11 *Metamorphosis III* (1967–1968) by Maurits Escher ©
1990 M. C. Escher Heirs/Cordon Art – Baarn – Holland.

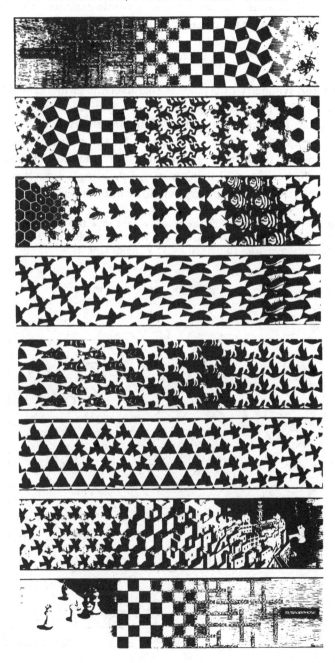

The result is a mosaic with a restructured pattern. It is only one example of how dynamic problems can help solve many difficult questions in the theory of tilings. Note that the potentialities of this method increase significantly if alongside the steady-state dynamic patterns we use meta-stable and quasi-stationary dynamic objects.

10.2 Phyllotaxis

The form and symmetry of crystals is characteristic of inorganic matter. In crystals we find the most primitive kind of organization, a monotonous repetition, i.e., the simple translation of an elementary cell. What is the degree of complexity of patterns necessary for the emergence of life?

Discussing the structure of genes, those giant molecules bearing the hereditary characteristics of living matter, Schrödinger did not consider periodic patterns, since they carry very little information [12]. He intro-

Fig. 10.1.12   The relief of the velocity field in a two-dimensional flow where the initial conditions and the external force field have different symmetries (obtained by V. V. Beloshapkin).

duced the concept of aperiodic crystals: regular patterns which contain much more information. Quasi-crystals represent the simplest form of such self-organization. A special part in the aperiodic organization of patterns is played by symmetry of the fifth order.

This symmetry is widespread among biological objects. Palaeontologists use it as a sure proof of the biological, rather than geological, origin of fossils. According to Belov [13], the axis of symmetry of the fifth order is a kind of mechanism in the struggle for existence among small organisms, a guarantee against petrification which would start with the trapping of a living organism in a lattice. Even the most primitive organisms not yet wholly belonging to organic matter, viruses, when exposed to X-ray analysis, display a high degree of structural regularity and fifth-order symmetry. Photographs made through an electron microscope show that several varieties of viruses resemble a regular icosahedron and that virus colonies form a structure with fifth-order symmetry. Other well-known examples of fifth-order symmetry in animals are the starfish (Fig. 10.2.1) and other denizens of the sea (Fig. 10.2.2). These animals display not only symmetry of a higher order but even, on occasion, the coexistence of different orders of symmetry [14]. Figure 10.2.3 borrowed from Haeckel's atlas [15], shows the coexistence of fourth and fifth-order symmetries. There are further examples of natural objects with quasi-symmetry: the shells of ammonite molluscs (Fig. 10.2.4) whose shape is, in fact, very close to certain plants. A certain amount of explanation, however, would be in order here.

Fig. 10.2.1   A starfish, showing fifth-order symmetry (Stevens [24]).

The structural organization in plants which manifests itself in the arrangement of flowers, seeds, scales, leaves, etc. (as in sunflowers, daisies, pine-cones and so on) is called phyllotaxis [8]. This phenomenon has been discussed by botanists for quite a long time. Usually, objects

Fig. 10.2.2   Echinoderms from the class of Ophiuroidea. From the Haeckel atlas [15].

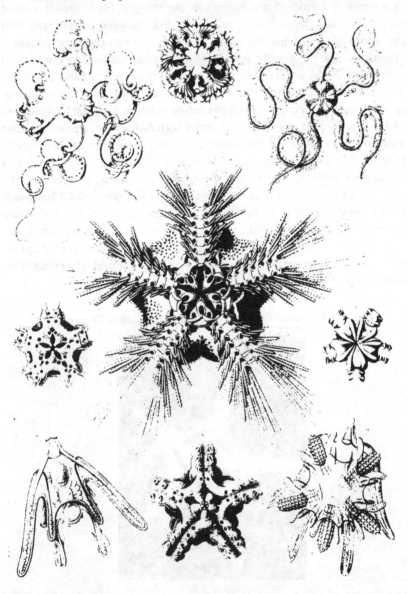

displaying phyllotaxis have a cylindrical or conical shape which has certain characteristics of a quasi-crystal with fifth-order symmetry [17].

It has been found [18, 19] that the seeds of a spruce cone or scales of a pineapple form a certain quasi-regular tiling of the surface in which we can easily notice the arrangement of the neighbouring cells in spirals

Fig. 10.2.3   An example of coexistence of symmetries of the fourth and fifth order (Melathallia) (Haeckel [15]).

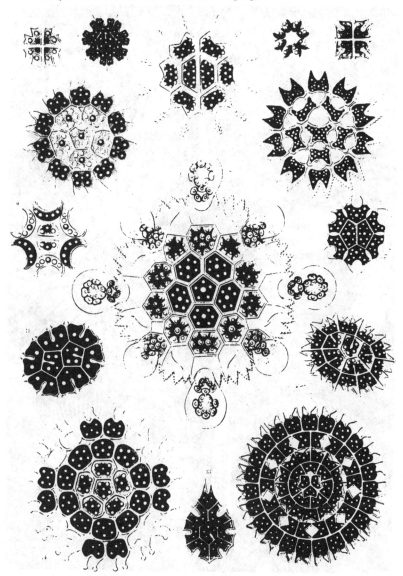

228 *Patterns in art and nature*

(parastichis). If a cell has a hexadecimal shape (as in the case of an aster or a pineapple), it belongs to three types of parastichis at once. But when the cell is rhombic in shape (e.g., pine-cones or sunflowers), there are only two types of parastichis – one clockwise, the other counterclockwise (Fig. 10.2.5). In most cases, for plants the numbers of parastichis of each

Fig. 10.2.4   The shells of ammonites (Ammonitida) – an example of selfsimilar logarithmic spirals (Haeckel [15]).

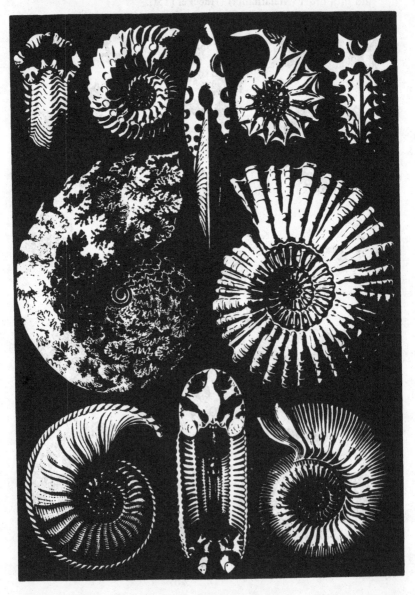

type are the successive Fibonacci numbers, their proportion being the appropriate ratio for the golden mean [20, 21] (Fig. 10.2.6).

If we dissect the surface of a pineapple and stretch it out, we get a strip lying between two parallel straight lines representing the initial vertical cut. In Fig. 10.2.7 these lines are $x = 0$ and $x = 1$, and the hexagonal cells are numbered in the order of increasing ordinates of their centres [22]. As the point of reference, we have taken the centre of the zero-order cell, while the centre of the cell with index '1' has the

Fig. 10.2.5   A sunflower head with partly removed seeds. The numbers of parastichis (34 and 55) are the Fibonacci numbers (Cook [19]).

coordinates (τ, h), where h is arbitrary and

$$\tau = \frac{\sqrt{5}-1}{2} \tag{10.2.1}$$

is the golden mean. The centre of the nth cell has the following coordinates [22]

$$y = nh; \qquad x = \frac{n}{\tau} - \left[\frac{n}{\tau}\right], \tag{10.2.2}$$

Fig. 10.2.6 Logarithmic spirals on the section of the apex of *Araucaria excelsa* (Church [21]). The numbers of parastichis are 8 and 13.

i.e., it is defined by the same quasi-crystal algorithm as, for example, the sequence of straight lines in the Ammann's lattice.

Another example of phyllotaxis is the shape of parastichis resembling a logarithmic spiral (Fig. 10.2.5), i.e., a curve possessing the property of selfsimilarity. This property is extremely important and is closely related to the properties of inflation and deflation in objects showing quasi-crystal symmetry. Thus, by simple partitioning or amalgamation of rhombi in the Penrose tiling, it can be transformed into a similar tiling consisting of rhombi of a greater size (inflation) or smaller size (deflation).

The property of selfsimilar transformation of the structure is programmed in plants by the genetic code; however, selfsimilarity should be inherent in the structures themselves. Its presence suggests that certain invariants that we observe on a macroscopic scale (for instance, the rotation number of the objects characterized by phyllotaxis, equal to the golden mean) must also remain on a microscopic scale on the molecular level.

Fig. 10.2.7 The surface of a pineapple stretched out on a plane (Coxeter [22]).

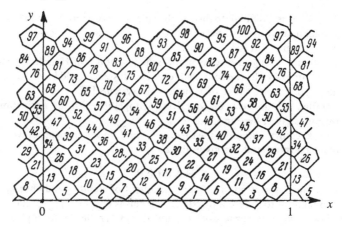

Notes

CHAPTER 1

1.1 Contemporary presentation of the Hamiltonian formalism with applications to numerous problems in physics is given in [1-3]. Some generalizations of the Hamiltonian formalism were triggered by development of exact integration techniques for nonlinear wave equations on the one hand, and by the analysis of non-integrable dynamic problems (constituting a majority) on the other [4, 5].

1.2 It is, of course, difficult to discuss the comparative properties of Hamiltonian and dissipative systems here. Many differences and their physical interpretations are given in [6].

1.3 We are not going to discuss non-dissipative and non-Hamiltonian systems here. For example, the so-called reversible systems could have many properties analogous to those of Hamiltonian systems [7].

1.4 Detailed solution can be found in [6] together with the derivation of all formulas given here in finite form.

1.5 The exact formulation of the theorem's conditions and its proof were carried out by Arnold and are given in [2].

CHAPTER 2

2.1 A simple description of different problems of stability and choas and their applications to different physical problems is given in [1]. Special problems of the chaos (dynamic stochasticity) theory can be found in [2-4].

2.2 More details on nonlinear resonance and its different applications are in [1, 2].

2.3 A presentation of the KAM theory, as well as many of the consequences deriving from this theory, can be found in [5, 6]. The Kolmogorov theory on the conservation of invariant tori was formulated in 1954. A proof was given by Arnold, while an analogous theorem was proved independently by Moser under somewhat different assumptions.

2.4 A more detailed presentation of this problem is given in [1]. Difficulties are encountered in trying to prove the property of mixing in the form of the exponential law (2.5.7) for $t \to \infty$. Real systems, as a rule, have not just one interval but many (even an infinite number of intervals is possible), each with different asymptotical behaviour for the decay of correlations. It is similarly difficult to establish the local instability in the form of the exponential behaviour (2.4.1) with an estimation of the value h_0 which is sufficiently uniform in time. However, there are reasonably convenient models, for which we can find h_0 and τ_c and check (2.5.8). These are, for example, Anosov's systems and certain types of billiards [2].

CHAPTER 3

3.1 The history of the discovery of the stochastic layer was as follows. In the early 1960s, physicists studying high-temperature plasma were actively engaged in investigation of the topology of magnetic surfaces for different configurations of magnetic traps. The stability of magnetic surfaces in toroidal systems was considered to be an indispensable condition for such devices. However, it should also be possible for magnetized particles to escape the trap by moving along magnetic field lines which do not belong to the invariant magnetic surfaces. It has been known since Poincaré's works, that motion near separatrices is unstable and trajectories behave in an extremely complicated way. The complexity was the result of a separatrix splitting under the effect of a perturbation. The estimation of the region of separatrix splitting was carried out in [3] in conjunction with the problem of stability of magnetic surfaces. In [4] the behaviour of magnetic surfaces was reduced to the problem of nonlinear resonances and their overlapping in equivalent dynamic systems. This allowed the introduction of the notion of chaos of magnetic field lines into the problem. In a subsequent paper [1] the region near the separatrix was specifically studied. The role of a time-dependent disturbance was played by a perturbation which was periodic along the axis of the torus. The same work includes an important numerical analysis which has shown the existence of local instability and spatial diffusion of magnetic field lines. Further generalization of these results in [2] proved the universality and indispensability of the stochastic layer. In many works one can often find some confusion around the notions of separatrix splitting and formation of the stochastic layer. In fact, these notions are not equivalent and their difference is not a trivial issue. In [5], for example, attention is drawn to the fact that separatrix splitting can take place even in an integrable case (i.e., in the absence of a stochastic layer). This paper also includes the opposite example, in which a stochastic layer exists in the vicinity of non-crossing separatrices.

3.2 If $\nu \lesssim \omega_0$, the overlapping of resonances takes place. In this case, almost the whole region inside separatrix loops becomes the region of stochastic motion. This leads to the fact that condition $\nu \lesssim \omega_0$ can be considered as the criterion of the onset of strong stochasticity (the Chirikov criterion of resonance overlapping [8]). At $\nu \lesssim \omega_0$, the modulation of the pendulum's frequency in equation (3.3.1) becomes adiabatic. Nevertheless, in this case also, almost the whole region inside separatrix loops is the region of chaos. However, in this case, mixing times are substantially greater [6, 7].

CHAPTER 4

4.1 Such models are Anosov systems [1], Sinai billiards [2] and so on. A simple description of these models can be found in [3].

4.2 Many aspects of these difficult problems are described in [3]. One of the interesting questions arising in conjunction with the potential coexistence of the region of chaos and the region of stability is what the topology of the chaos boundary is.

4.3 On fractals, fractal objects and their examples in nature see Mandelbrot [4]. Various manifestations of fractality of chaos are discussed in [5, 6]. An accessible description of the interrelation of fractals with chaos is given in [3]. Chaos in Fig. 4.0.1 is related to the class of the so-called 'fat fractals' [7].

4.4 Long before Green's work there were approximate qualitative approaches to define K_c. First of all, one should mention the Chirikov method of resonance overlapping [9, 10]. Later, an estimate of K_c was given which was based on the condition of local phase instability [11, 12]. Both approaches gave $K_c \sim 1$. Then, some attempts were made to determine the boundary of the stochastic sea with better accuracy [10, 13]. All of them were based on the use of a certain (and rather strong) extrapolation of results from the region of small values of K to the region of $K \sim 1$ [14].

4.5 Since a rigorous technique to define the boundary of global chaos in the form (4.1.11) or (4.1.12) is lacking, we cannot list precisely all the necessary conditions for the derivation of these formulas. Therefore, it is impossible, a priori, to exclude other possibilities even in cases when the KAM theory is applicable.

4.6 The analysis of transport processes in the presence of Cantori is carried out in [21, 22]. One can conclude that their presence not only represents a considerable interest for plasma physics applications but creates difficulties in the rigorous theory of chaos systems. Indeed, barriers are unavoidable, which cause anomalously long wandering of trajectories along them. In the distribution function of such time intervals of delay one is likely to find arbitrarily large intervals of delay. This means that in a trajectory which must be chaotic there are regular insets, in which the system spends a long time. This very property of the motion leads to a weak (power-like) damping of correlators [23].

4.7 The average velocity of particles moving in chaotic fields can increase. This acceleration mechanism was suggested by Fermi to explain the origin of cosmic rays [24]. Later Ulam suggested a seemingly simple model for analysis in which the particle is moving between two walls, one of which is periodically oscillating [25]. Numerical and analytical analysis of Ulam's model in [26], proved the existence, in such a model, of chaotic dynamics and stochastic acceleration of particles. Later these studies were continued in many works (see the review of their results in [3, 10, 12, 27]). Now there are newer, more realistic models of a particle's acceleration which were originally suggested in plasma physics [28–20].

4.8 There is a substantial literature on the phenomenon of intermittency. Intermittency is directly related to the multi-fractality of the process, i.e., to a non-uniform distribution of singular or fractal properties in space or in time. We shall not discuss this problem here since its applications are very rich and require additional information. Instead, we provide references to the most complete reviews [33–35].

4.9 Strong electric fields of relativistic ($\omega/k \sim c$) potential waves in plasma may be used to accelerate particles up to superhigh energies [37, 38]. One of the best studied acceleration schemes is the so called beat-wave accelerator

[39]. In this, a particle is accelerated by a plasma wave generated by two collinear laser beams in a non-uniform plasma. When emission by one of the laser beams is broad-band, the electric field $E(x, t)$, induced in the plasma, also has a large spectral width.

The results presented in this section are discussed in [40].

CHAPTER 5

5.1 There are several examples of Arnold diffusion in problems concerning plasma physics and dynamics of particles in accelerators (a review is given in [2–4]). An estimate of the the web's width $\sim \exp(-\text{const}/\sqrt{\varepsilon})$ is given in [5]. The velocity of diffusion is estimated in [6]. The diffusion on the web has its own specifics which is discussed somewhat later.

5.2 Equation (5.2.8) describes the wave–particle interaction in a constant magnetic field. This problem has numerous applications in plasma physics (reviews [3, 12]). A problem of wave propagation obliquely to a magnetic field was first considered in [13]. Chaotization at the perpendicular propagation of the wave was discovered in [14–16]. This problem appeared to be very rich in different physical situations and to date there are still some unexplored or not sufficiently studied cases. See also the analysis of equation (5.2.8) in conjunction with the paradox of Landau damping disappearance [17–20].

5.3 Solutions (5.3.13), (5.3.16) and so on for the motion inside a square cell were published in [14]. Later, they were used in [15, 16]. The asymptotics of large values allow us to reach conclusions on the local properties of the phase portrait of the system (5.2.3).

5.4 This problem was studied in [10, 11].

CHAPTER 6

6.1 A more detailed analysis of the mechanism of particle acceleration on the wave front is given in conjunction with numerous applications to plasma physics in [3].

6.2 In [5–7] the period-doubling bifurcations were studied in maps preserving the phase volume. Numerical analysis gave a constant $\delta = 8.72109720$. One can expect that this value is universal for Hamiltonian systems.

6.3 Fractals represent sets organized in too complex a way to be simply called lines, planes and so on. A detailed but accessible presentation is given in Mandelbrot [8]. Some idea of Koch's fractal, can be obtained in the following way. Let us consider an element of a saw-tooth with a certain number of teeth (e.g., three). Each side of each tooth we can transform into a saw-tooth again, but with a smaller size of teeth. If we continued this process, always converting the size of decreasing teeth in the saw-tooth of a smaller size, eventually we would obtain a complicated curve with a finite number of teeth. If continued to infinity, this process would lead to an example of Koch's fractal.

6.4 Acceleration of relativistic particles in a magnetic field due to the existence of a web has numerous applications. Some of them are discussed in [14, 15].

6.5 Saturation of the web can take place not only because of the nonlinearity of undisturbed motion but also in the case of a perturbation represented by a finite number of harmonics [16].

CHAPTER 7

7.1 The famous studies of Kepler on the form of snowflakes (1611) [1] and in *Harmonice mundi* (1619) [2] have triggered many attempts to penetrate into the geometric mysteries of the world's structure. Analysing the potential causes of the hexagonal shape of a snowflake and the polygonal (instead of spherical) shape of pomegranate grains, Kepler tried to identify the forces which can lead to such structures. In the same way, he tried to explain the sizes of orbits of the then known planets with the help of geometric constructions based on the use of Plato's regular objects. The idea of dependence of the geometric forms of objects on the types of dynamic laws which control these objects was widely discussed later by Hermann Weyl [3].

7.2 There is a unique book [4] in which a large number of different types of plane tilings is collected. It also gives contemporary results from the theory of two-dimensional tilings. A study of geometrical properties of mosaics is also given in [5–8].

7.3 Penrose published his versions of pentagonal tilings in 1974 [11]. However, detailed study started only after the appearance of the paper by Gardner [12] which attracted the attention of wide circles of researchers not only to Penrose's work but to the problems of symmetry of the fifth order in general. Immediately a serious attempt was made to use the Penrose tiling as a possible model of a crystal with fifth-order symmetry [13]. Later, systematic searches led to the discovery of a quasi-crystal which has rotational symmetry of the fifth order at the absence of translational symmetry [14]. A large number of papers appeared on the theoretical and experimental analysis of structures of crystals with anomalous symmetries (quasi-crystals) [15–17]. Substantial attention was devoted to the geometry of Penrose tilings in [4]. And although the question of the real existence of quasi-crystals is not completely clarified, quasi-crystal-type tilings and symmetries have been found to exist in some real dynamic systems (see review [21]).

7.4 At the basis of the projection technique is the method suggested in [22] and later developed in many works [23–28, 16, 17].

7.5 The relationship between generator $\hat{M}_5$ and Markovian partitionings of a plane was mentioned by Arnold [29].

7.6 Strictly speaking, the boundary of Hamiltonian chaos is still only poorly studied. Dynamic properties in its neighbourhood and properties of the phase portrait in the vicinity of this boundary are still awaiting explanation. The complexity of the trajectories of the systems in the vicinity of the boundary of chaos is the result of a large number of small islands of stability in which the stochastic trajectory cannot penetrate. Thus, it circumvents it and the final result becomes non-trivial.

7.7 Different ideas of decoration can be found in [4]. Discussion of some particular questions is given in reviews [21, 30].

7.8 Just as the golden mean τ_0 is the solution of a quadratic equation satisfied by the value $\cos(2\pi/5)$, the values τ_1, τ_2 are the solutions of cubic equations for $\cos(2\pi/7)$. These numbers are decomposed in branching continuous fractions.

7.9 The fundamental analysis of the notion of local isomorphism is given in [4]. See also [17, 32].

7.10 The Fourier spectrum of the Penrose tiling was given in [34], somewhat before an analogous result was obtained experimentally [14]. Intercomparison of Fourier spectra of the patterns generated by the Hamiltonian H_5, and the experimental ones, was carried out in [35]. All these data allow us to believe that the Fourier spectrum of a plane tiling reflects the symmetry properties of the tiling sufficiently well, though it does not allow the unequivocal reconstruction of the pattern.

7.11 In fact, singular trajectories are also present in the case of strong chaos. They are related to many interesting geometric properties of dynamic systems, on the one hand, and to some problems in number theory, on the other [37, 38].

7.12 An analogous paradoxical coexistence of the properties of ordered and disordered patterns is seen in electron spectra of two-dimensional patterns in two-dimensional models with pentagonal symmetry [39, 40].

CHAPTER 8

8.1 Only comparatively recently have the studies of the origin of patterns in a continuous medium reached quite a high level. To typical objects in such studies one can relate thermal convection, electrodynamic convection, mixing due to the Raleigh–Taylor instability and so on. Such structures are very well defined visually and have a good (regular) geometry. Experimentally, one can study their generation, evolution and interaction. There are several books which contain detailed material on these issues [7, 8].

8.2 A large number of examples of two-dimensional steady-state vortex flows of unbounded fluid are given in [3, 11]. With the help of the method of variables separation, a three-parametrical class of solutions of equation (8.1.10) can be found in [12], with $f(\Psi) = \Psi \ln \Psi + a\Psi$ and $f(\Psi) = \Psi^b$ where a and b are real numbers. In [13] also, the representation of vorticity in the form $f(\Psi) = \sin \Psi$ and $f(\Psi) = a\,e^{-2\Psi} + e^{\Psi}$ is considered.

8.3 Recently, hydrodynamic flows of the type of (8.1.12) and (8.1.14) have fallen under consideration in conjunction with different problems in hydrodynamics [15–20]. It is likely that with increasing experimental opportunities, attention to the patterns of quasi-crystal type will grow, since they can originate as a result of subsequent bifurcations in the process of transition from laminar to turbulent flow.

8.4 The problem of stability of laminar flow (8.2.4) was raised by Kolmogorov at his seminar [22]. The linear theory of stability for this flow was given in [21]. In [23] the proof was given, at $\text{Re} > \text{Re}_c$, that secondary steady-state flows are stable relative to the periodic perturbations with the period of the principle flow. However, in [24–26] it was proved that all secondary flows with translational symmetry are unstable against a perturbation with a period incommensurate with the period of the principal flow. Thus, the theoretical conclusion was reached that, in the over-critical region at $\text{Re} > \text{Re}_c$, space-periodic secondary flows are unstable. However, the experiments discussed in [27] displayed secondary stationary patterns in the form of a regular lattice of vortices arranged as on a chessboard and changing its shape with the increase of the Reynolds number. Explanation of this paradox has to do with friction at the floor. The corresponding theory allowing a quantitative description of the experimental results was given in [27, 28].

CHAPTER 9

9.1   There is a complete analogy between the behaviour of stream lines and magnetic field lines (and surfaces). Magnetic surfaces are invariant spatial patterns in which the family of magnetic field lines is lying. The problem of the existence of magnetic surfaces is equivalent to the problem of the existence of the invariant tori of a dynamic system. The destruction of magnetic surfaces is manifested in chaotic behaviour of magnetic field lines, although the corresponding spatial configuration of the magnetic field is regular. The Hamiltonian formalism was applied to the problem of magnetic surfaces in [1, 2]. The subsequent development of this topic is described in [3, 4].

9.2   The study of stochastic dynamics of passive scalar impurity (advection) in a non-steady-state two-dimensional flow was carried out in [8]. According to Lagrangian description, this problem is reduced to the study of chaotic behaviour of the solution of a non-autonomous Hamiltonian system. The experimental study of the mixing of passive impurity in a fluid between two non-coaxially rotating cylinders is described in [9, 10].

9.3   Numerical analysis of the behaviour of solutions of the induction equations (9.1.11), describing the evolution of a magnetic field in a steady-state flow of a conducting fluid was initiated in [15] and later followed in [16, 17]. For these studies the velocity field of the ABC-flow was chosen with the coefficients $A = B = C = 1$. In [15] it was discovered that a dynamo effect takes place for the magnetic Reynolds numbers $9 < R_m < 17.5$. In [17] numerical computations were upgraded to $R_m = 450$, and the second region of the existence of a hydrodynamic dynamo $R_m > 27$ was discovered. In [18] it was shown that a magnetic field generated in this way has a tendency to concentrate on fractals.

References

CHAPTER 1

1 ter-Haar, D., *Elements of Hamiltonian Mechanics* (Pergamon Press, Oxford, 1971).

2 Arnold, V. I., *Mathematical Methods in Classical Mechanics* (Springer, New York, 1978).

3 Dubrovin, B. A., S. P. Novikov and A. T. Fomenko, *Modern Geometry* (Moscow, Nauka, 1979).

4 Dubrovin, B. A., I. M. Krichever and S. P. Novikov, Integrable systems 1. In *Encyclopaedia of Mathematical Sciences*, v. 4 (Springer, New York, 1988).

5 Arnold, V. I., V. V. Kozlov and A. I. Neishtadt, Mathematical aspects of classical and celestial mechanics. In *Encyclopaedia of Mathematical Sciences*, v. 3 (Springer, New York, 1988).

6 Sagdeev, R. Z., D. A. Usikov and G. M. Zaslavsky, *Nonlinear Physics* (Harwood Academic Publishers, New York, 1988).

7 Arnold, V. I. and M. B. Sevryuk. In *Nonlinear Phenomena in Plasma Physics and Hydrodynamics*, ed. R. Z. Sagdeev (MIR Publ., Moscow, 1986, p. 31).

8 Gradshtein, I. S. and I. M. Ryzhik, *Tables of Integrals, Sums, Series and Products* (Moscow, Fizmatgiz, 1963).

CHAPTER 2

1 Sagdeev, R. Z., D. A. Usikov and G. M. Zaslavsky, *Nonlinear Physics* (Harwood Academic Publishers, New York, 1988).

2 Zaslavsky, G. M., *Chaos in Dynamic Systems* (Harwood Academic Publishers, New York, 1985).

3 Lichtenberg, A. and M. Liberman, *Regular and Stochastic Motion* (Springer, New York, 1983).

4 Chirikov, B. V., *Phys. Rep.*, **52**, 263 (1979).

5   Arnold, V. I., *Mathematical Methods in Classical Mechanics* (Springer, New York, 1978).

6   Arnold, V. I., V. V. Kozlov and A. I. Neishtadt, Mathematical aspects of classical and celestial mechanics. In *Encyclopaedia of Mathematical Sciences*, v. 3 (Springer, New York, 1988).

7   Arnold, V. I., *Dokl. Akad. Nauk SSSR*, **156**, 9 (1964).

8   Kolmogorov, A. N., *Dokl. Akad. Nauk SSSR*, **119**, 861 (1958); **124**, 754 (1959).

9   Sinai, Ya. G., *Dokl. Akad. Nauk SSSR*, **124**, 768 (1959); **125**, 1200 (1959).

CHAPTER 3

1   Filonenko, N. N., R. Z. Sagdeev and G. M. Zaslavsky, *Nucl. Fusion*, **7**, 253 (1967).

2   Zaslavsky, G. M. and N. N. Filonenko, *Zh. Eksp. Teor. Fiz.*, **54**, 1590 (1968).

3   Melnikov, V. K., *Dokl. Akad. Nauk SSSR*, **148**, 1257 (1963).

4   Rosenbluth, M. N., R. Z. Sagdeev, J. B. Taylor and G. M. Zaslavsky, *Nucl. Fusion*, **6**, 297 (1966).

5   Ziglin, S. L., *Proc. Moscow Math. Soc.*, **41**, 287 (1980).

6   Menyuk, C. R., *Phys. Rev.*, **A31**, 3282 (1985).

7   Cary, J. R., D. F. Escande and J. L. Tennyson, *Phys. Rev.*, **A34**, 4256 (1986).

8   Chirikov, B. V., *Phys. Rep.*, **52**, 263 (1979).

9   Sagdeev, R. Z., D. A. Usikov and G. M. Zaslavsky, *Nonlinear Physics* (Harwood Academic Publishers, New York, 1988).

10 Zaslavsky, G. M., *Chaos in Dynamic Systems* (Harwood Academic Publishers, New York, 1988).

11 Lichtenberg, A. and M. Liberman, *Regular and Stochastic Motion* (Springer, New York, 1983).

12 Ulam, S. M., *A Collection of Mathematical Problems* (Interscience Publishers, New York, 1960).

13 Wisdom, J., S. J. Peale and J. Magnard, *Icarus*, **58**, 137 (1984).

14 Wisdom, J., *Icarus*, **72**, 241 (1987).

15 Wisdom, J., *Astron. J.*, **94**, 1350 (1987).

CHAPTER 4

1   Anosov, D. V., in *Proceedings of the Steklov Institute of Mathematics* (American Mathematical Society, Providence, Rhode Island, 1969).

2   Sinai, Ya. G., *Dokl. Akad. Nauk SSSR*, **153**, 1261 (1963).

3   Sagdeev, R. Z., D. A. Usikov and G. M. Zaslavsky, *Nonlinear Physics* (Harwood Academic Publishers, New York, 1988).

4   Mandelbrot, B., *The Fractal Geometry of Nature* (Freeman, San Francisco, 1982).

5   *Dimensions and Entropies in Chaotic Systems*, ed. G. Mayer-Kress (Springer, Berlin, 1986).

6   Pietronero L. and I. Tosati, *Fractals in Physics* (North-Holland, Amsterdam, 1986).

7   Umberger, D. K. and D. Farmer, *Phys. Rev. Lett.*, **55**, 661 (1985).

8   Green, J. M., *J. Math. Phys.*, **20**, 1183 (1979).

9   Chirikov, B. V., *Atomic Energy*, **6**, 630 (1959).

10   Chirikov, B. V., *Phys. Rep.*, **52**, 263 (1979).

11   Filonenko, N. N., R. Z. Sagdeev and G. M. Zaslavsky, *Nucl. Fusion*, **7**, 253 (1985).

12   Zaslavsky, G. M., *Chaos in Dynamic Systems* (Harwood Academic Publishers, New York, 1985).

13   Escande, D. F., *Phys. Rep.*, **121**, 163 (1986).

14   Arnold, V. I., *Geometric Methods in the Theory of Ordinary Differential Equations* (Springer, New York, 1983).

15   Percival, I. C., *J. Phys. A*, **7**, 794 (1974).

16   Percival, I. C., *J. Phys. A*, **12**, L57 (1979).

17   Percival, I. C. In: *Nonlinear Dynamics and the Beam–Beam Interaction*, ed. Month, Herrera (Amer. Inst. of Physics, Conf. Proceed., New York) **57**, 302 (1979).

18   Landau, L. D. and E. M. Lifshitz, *Mechanics* (Pergamon, Oxford, 1976).

19   Aubry, S., *Physica*, **7D**, 240 (1983).

20   Mather, J. N., *Topology*, **21**, 457 (1982).

21   MacKay, R. S., J. D. Meiss and I. C. Percival, *Physica*, **13D**, 55 (1984).

22   Percival, I. C. In: *Dynamical Chaos*, Proc. of a Roy. Soc. Meeting. London. The Royal Society, 131 (1987).

23   Karney, C. F. F., *Physica*, **8D**, 360 (1983).

24   Fermi, E., *Phys. Rev.*, **75**, 1169 (1949).

25   Ulam, S., *Proc. 4th Berkeley Symp. on Math. and Prob. - Berkeley - Los Angeles*, **3**, 315 (1961).

26   Zaslavsky, G. M. and B. V. Chirikov, *Dokl. Akad. Nauk SSSR*, **159**, 306 (1964).

27   Lichtenberg, A. J. and M. A. Lieberman, *Regular and Stochastic Motion* (Springer, New York, 1983).

28   Zaslavsky, G. M., *Zh. Eksp. Teor. Fiz.*, **88**, 1984 (1985).

29   Zaslavsky, G. M. and A. A. Chernikov, *Zh. Eksp. Teor. Fiz.*, **89**, 1632 (1985).

30   Berzin, A. A., G. M. Zaslavsky, S. S. Moiseev and A. A. Chernikov, *Fizika Plasmy*, **13**, 592 (1987).

31   Batchelor, G. K. and A. A. Townsend, *Proc. Roy. Soc. A.*, **199**, 238 (1949).

32   Pomeau, Y. and P. Manneville, *Comm. Math. Phys.*, **79**, 149 (1980).

33   Paladin, G. and A. Vulpiani, *Phys. Rep.*, **156**, 148 (1987).

34   Zeldovich, Ya. B., S. A. Molchanov, A. A. Ruzmaikin and D. D. Sokolov, *Usp. Fiz. Nauk*, **152**, 3 (1987).

35   Stanley, H. E. and P. Meakin, *Nature*, **335**, 405 (1988).

36   Zaslavsky, G. M., M. A. Malkov, R. Z. Sagdeev and A. A. Chernikov, *Fizika Plasmy*, **14**, 307 (1988).

37   Tajima, T. and J. M. Dawson, *Phys. Rev. Lett.*, **43**, 267 (1979).

38   Katsouleas, T. and J. M. Dawson, *Phys. Rev. Lett.*, **51**, 392 (1983).

39   Tajima, T., Physics of high energy particle accelerators. In *AIP Conference Proceedings. 127*, ed. M. Month, P. Dahe and M. Dienes (AIP, New York, 1985), p. 793.

40   Chernikov, A. A., T. Tel, G. Vattay and G. M. Zaslavsky, *Phys. Rev.*, **A40**, 4072 (1989).

41   Landau, L. D. and E. M. Lifshitz, *Field Theory* (Pergamon, Oxford, 1976).

CHAPTER 5

1   Arnold, V. I., *Dokl. Akad. Nauk SSSR*, **156**, 9 (1964).

2   Chirikov, B. V., *Phys. Rep.*, **52**, 263 (1979).

3   Lichtenberg, A. and M. Liberman, *Regular and Stochastic Motion* (Springer, New York, 1983).

4   Arnold, V. I., V. V. Kozlov and A. I. Neishtadt, Mathematical aspects of classical and celestial mechanics. In *Encyclopaedia of Mathematical Sciences*, v. 3 (Springer, New York, 1988).

5   Zaslavsky, G. M., *Chaos in Dynamic Systems* (Harwood Academic Publishers, New York, 1985).

6   Nekhoroshev, N. N., *Usp. Mat. Nauk*, **32**, 5 (1977).

7   Zaslavsky, G. M., R. Z. Sagdeev, D. A. Usikov and A. A. Chernikov, *Usp. Fiz. Nauk*, **156**, 193 (1988).

8   Zaslavsky, G. M., M. Yu. Zakharov, R. Z. Sagdeev, D. A. Usikov and A. A. Chernikov, *Zh. Eksp. Teor. Fiz.*, **91**, 500 (1986).

9   Chernikov, A. A., R. Z. Sagdeev, D. A. Usikov, M. Yu. Zakharov and G. M. Zaslavsky, *Nature*, **326**, 559 (1987).

10   Chernikov, A. A., M. Ya. Natenzon, B. A. Petrovichev, R. Z. Sagdeev and G. M. Zaslavsky, *Phys. Lett.*, **122A**, 39 (1987).

11   Chernikov, A. A., M. Ya. Natenzon, B. A. Petrovichev, R. Z. Sagdeev and G. M. Zaslavsky, *Phys. Lett.*, **129A**, 377 (1988).

12 Sagdeev, R. Z. and G. M. Zaslavsky, Regular and chaotic dynamics of particles in a magnetic field. In: *Nonlinear Phenomena in Plasma Physics and Hydrodynamics*, ed. R. Z. Sagdeev (Mir, Moscow, 1986).

13 Smith, G. R. and A. N. Kaufman, *Phys. Rev. Lett.*, **34**, 1613 (1975).

14 Fukuyama, A., H. Momota, R. Itatani and T. Takizuka, *Phys. Rev. Lett.*, **38**, 701 (1977).

15 Karney, C. F. F., *Phys. Fluids*, **21**, 1584 (1978).

16 Karney, C. F. F., *Phys. Fluids*, **22**, 2188 (1979).

17 Sagdeev, R. Z. and V. D. Shapiro, *Pis'ma Zh. Eksp. Teor. Fiz.*, **17**, 389 (1973).

18 Malkov, M. A. and G. M. Zaslavsky, *Phys. Lett.*, **105A**, 257 (1984).

19 Zaslavsky, G. M., M. A. Malkov, R. Z. Sagdeev and V. D. Shapiro, *Fizika Plasmy*, **12**, 788 (1986).

20 Zaslavsky, G. M., A. I. Neishtadt, B. A. Petrovichev and R. Z. Sagdeev, *Fizika Plasmy*, **15**, 631 (1989).

CHAPTER 6

1 Zaslavsky, G. M., M. Yu. Zakharov, R. Z. Sagdeev, D. A. Usikov and A. A. Chernikov, *Zh. Eksp. Teor. Fiz.*, **91**, 500 (1986).

2 Chernikov, A. A., R. Z. Sagdeev, D. A. Usikov, M. Yu. Zakharov and G. M. Zaslavsky, *Nature*, **326**, 559 (1987).

3 Zaslavsky, G. M., R. Z. Sagdeev, D. A. Usikov and A. A. Chernikov, *Usp. Fiz. Nauk*, **156**, 193 (1988).

4 Arnold, V. I., *Usp. Mat. Nauk*, **42**, 139 (1987).

5 Green, J. M., R. S. Mackay, F. Vivaldi and M. J. Feigenbaum, *Physica*, **3D**, 468 (1981).

6 Collet, P., J. P. Eckman and H. Koch., *ibid.* p. 457.

7 Bountis, T. C., *ibid.* p. 557.

8 Mandelbrot, B., *The Fractal Geometry of Nature* (Freeman, San Francisco, 1982).

9 Landau, L. D. and E. M. Lifshitz, *Statistical Physics* (Pergamon, Oxford, 1976).

10 Weyl, H., *Symmetry* (Princeton University Press, Princeton, 1952).

11 Zaslavsky, G. M., M. Yu. Zakharov, R. Z. Sagdeev, D. A. Usikov and A. A. Chernikov, *Pis'ma Zh. Eksp. Teor. Fiz.*, **44**, 349 (1986).

12 Lichtenberg, A. J. and B. P. Wood, *Phys. Rev.*, **A39**, 2153 (1989).

13 Longcope, D. W. and R. N. Sudan, *Phys. Rev. Lett.*, **59**, 1500 (1987).

14 Carioli, S. M. *Nuovo Cimento*, 1988.

15 Karimabadi, H. and V. Angelopoulos, *Phys. Rev.*, **62**, 2342 (1989).

16 Murakami, S., T. Sato and A. Hasegawa, *Physica*, **32D**, 269 (1988).

References

1   Kepleris, J. *Strena, seu de nive sexangula. Francofurti ad Moemum: apud Tampach*, 1611, 21p.
2   Kepler, J., *Weltharmonid* (Munich, Oldenburg, 1967).
3   Weyl, H., *Symmetry* (Princeton University Press, Princeton, 1952).
4   Grunbaum, B. and G. C. Shepard, *Tilings and Patterns* (Freeman, New York, 1987).
5   *The Mathematical Gardner*, ed. D. A. Klarner (Prindle, Weber and Schmidt, Boston, 1981).
6   *Patterns of Symmetry*, ed. M. Senechal and H. Fleck (University of Massachusetts Press, Amherst, 1977).
7   Koptsig, V. A. and A. V. Shubnikov, *Symmetry in Art and Science* (Nauka, Moscow, 1972).
8   Coxeter, H. S. M., *Introduction to Geometry* (Wiley, New York, 1961).
9   *The World of M. C. Escher*, ed. J. L. Locher (Abrahams, New York, 1977).
10   Schrödinger, E., *What is Life?* (Cambridge University Press, Cambridge, 1944).
11   Penrose, R., *Bull. Inst. Math. Appl.*, **10**, 266 (1974).
12   Gardner, M., *Scient. Amer.*, **236**, 110 (1977).
13   MacKay, A. L., *Sov. Phys. Crystallogr.*, **26**, 517 (1981).
14   Schechtman, D., I. Blech, K. Gratias and J. W. Cahn, *Phys. Rev. Lett.*, **53**, 1951 (1984).
15   Int. Workshop on Aperiodical Crystals, *J. de Physique*, **47**, Coll. C3, Suppl. 7 (1986).
16   Levine, D. and P. J. Steinhardt, *Phys. Rev. Lett.*, **53**, 2477 (1984).
17   Korepin, V. E., *Quasiperiodic Tilings and Quasicrystals, Proceedings of Scientific Seminars LOMI*, **155**, 116 (Nauka, Leningrad, 1986).
18   Chernikov, A. A., R. Z. Sagdeev, D. A. Usikov and G. M. Zaslavsky, *Phys. Lett.*, **125A**, 101 (1987).
19   Chernikov, A. A., R. Z. Sagdeev, D. A. Usikov, M. Yu. Zakharov and G. M. Zaslavsky, *Nature*, **326**, 559 (1987).
20   Zaslavsky, G. M., M. Yu. Zakharov, R. Z. Sagdeev, D. A. Usikov and A. A. Chernikov, *Pis'ma Zh. Eksp. Teor. Fiz.*, **44**, 349 (1986).
21   Zaslavsky, G. M., R. Z. Sagdeev, D. A. Usikov and A. A. Chernikov, *Usp. Fiz. Nauk.*, **156**, 193 (1988).
22   De Bruijn, N. G., *Kon. Bederl. Akad. Wetesch. Proc., Ser.* **A84**, 38, 53 (1981).
23   Kramer, P. and R. Neri, *Acta Crystallogr., Sec.* **A40**, 580 (1984).
24   Elser, M., *Phys. Rev.*, **32B**, 4892 (1985).

25 Kalugin, P. A., A. Yu. Kitaev and L. S. Levitov, *Pis'ma Zh. Eksp. Teor. Fiz.*, **41**, 119 (1985).

26 Duneau, M. and A. Katz, *Phys. Rev. Lett.*, **54**, 2688 (1985).

27 De Bruijn, N. G., *J. de Physique*, **47**, Coll. C3, Suppl. 7, 9 (1986).

28 Gahler, F. and T. Rhyner, *J. Phys.*, **19A**, 267 (1986).

29 Arnold, V. I., *Usp. Mat. Nauk*, **42**, 139 (1987).

30 Chernikov, A. A., R. Z. Sagdeev, D. A. Usikov and G. M. Zaslavsky, *Computers Math. Applic.*, **17**, 17 (1989).

31 Levine, D. See [15], p. 125.

32 Levine, D. and P. J. Steinhardt, *Phys. Rev.*, **84B**, 596 (1986).

33 Choy, T. C., *Phys. Rev. Lett.*, **55**, 2915 (1985).

34 MacKay, A. L., *Physica*, **114A**, 609 (1982).

35 Lubensky, J. C., J. E. S. Socolar, P. J. Steinhardt, P. A. Bancel and P. A. Heiney, *Phys. Rev. Lett.*, **57**, 1440 (1986).

36 Landau, L. D. and E. M. Lifshitz, *Statistical Physics* (Pergamon, Oxford, 1977).

37 Percival, I. C. and F. Vivaldi, *Physica*, **25D**, 105 (1987).

38 Vivaldi, F., *Proc. R. Soc., Lond.*, **413A**, 97 (1987).

39 Ueda, K. and H. Tsunetsugu, *Phys. Rev. Lett.*, **58**, 1272 (1987).

40 Kuzmenko, A. V., I. R. Sagdeev, D. A. Usikov and R. Z. Sagdeev, *Phys. Lett.*, **130A**, 381 (1988).

41 Kornfeld, I. P., Ya. G. Sinai and S. V. Fomin, *Ergodic Theory* (Nauka, Moscow, 1980).

CHAPTER 8

1 Karman, T., *Gott. Nachr., Math.-Phys. Kl.* (1912) p. 547.

2 Villat, H., *Leçons sur la Théorie des Tourbillons* (Paris, Gauthier-Villars et C°, 1930).

3 Lamb, H., *Hydrodynamics* (Cambridge University Press, Cambridge, 1932).

4 Benard, H., *Revue générale des sciences pures et appliquées*, **11**, 1261 (1900).

5 Rayleigh, Lord., *Phil. Mag.*, **32**, 529 (1916).

6a Koschmieder, E. L. and M. I. Biggerstaff, *J. Fluid Mech.*, **167**, 49 (1986).

6b Le Gal, P., A. Pocheau and V. Croquette, *Phys. Rev. Lett.*, **54**, 2501 (1985).

7 *Cellular Structures in Instabilities*, ed. J. E. Wesfreid and S. Zaleski (Springer, Berlin, 1984).

8 *Propagation in Systems Far from Equilibrium*, ed. J. E. Wesfreid, H. R. Brand, P. Manneville, H. Albinet and N. Boccara (Springer, Berlin, 1988).

9 *New Trends in Nonlinear Dynamics and Pattern Forming Phenomena: The Geometry of non Equilibrium*, ed. P. Huerre and P. Coullet (Plenum Press, New York, 1988).

10 Landau, L. D. and E. M. Lifshitz, *Fluid Mechanics* (Pergamon Press, Oxford, 1959).

11 Batchelor, G. K., *An Introduction to Fluid Dynamics* (Cambridge University Press, Cambridge, 1970).

12 Shercliff, J. A., *J. Fluid Mech.*, **82**, 687 (1977).

13 Kaptsov, O. V., *Dokl. Akad. Nauk SSSR*, **298**, 597 (1988).

14 Stuart, J. T., *J. Fluid Mech.*, **29**, 417 (1967).

15 Bayly, B. J. and V. Yakhot, *Phys. Rev.*, **A34**, 381 (1986).

16 Zaslavsky, G. M., R. Z. Sagdeev and A. A. Chernikov, *Zh. Eksp. Teor. Fiz.*, **94**, 102 (1988).

17 Beloshapkin, V. V., A. A. Chernikov, M. Ya. Natenzon, B. A. Petrovichev, R. Z. Sagdeev and G. M. Zaslavsky, *Nature*, **337**, 133 (1989).

18 Beloshapkin, V. V., A. A. Chernikov, R. Z. Sagdeev and G. M. Zaslavsky, *Phys. Lett.*, **133A**, 395 (1988).

19 Yakhot, V., B. J. Bayly and S. A. Orszag, *Phys. Fluids*, **29**, 2025 (1986).

20 Malomed, B. A., A. A. Nepomnyashchii and M. I. Tribelsky, *Pis'ma Zh. Tekhn. Fiz.*, **13**, 1165 (1987).

21 Meshalkin, L. D. and Ya. G. Sinai, *Prikl. Matem. and Mekhan.*, **25**, 1140 (1961).

22 Arnold, V. I. and L. D. Meshalkin, *Usp. Mat. Nauk*, **15**, 247 (1960).

23 Yudovich, V. I., *Prikl. Matem. and Mekhan.*, **29**, 453 (1965).

24 Nepomnyashchii, A. A., *Pricl. Matem. and Mekhan.*, **40**, 886 (1976).

25 Klyatskin, V. I., *Prikl. Matem and Mekhan.*, **36**, 263 (1972).

26 Green, J. S. A., *J. Fluid Mech.*, **62**, 273 (1974).

27 Bondarenko, N. F., M. Z. Gak and F. V. Dolzhansky, *Izv. Akad. Nauk SSSR, Fizika Atm. i Okeana*, **15**, 1017 (1979).

28 Gledzer, E. B., F. V. Dolzhansky and A. V. Obukhov, *Systems of Hydrodynamical Types and Their Use* (Moscow, Nauka, 1981).

29 Sivashinsky, G. and V. Yakhot, *Phys. Fluids*, **28**, 1040 (1985).

CHAPTER 9

1 Rosenbluth, M. N., R. Z. Sagdeev, J. B. Taylor and G. M. Zaslavsky, *Nucl. Fusion*, **6**, 297 (1966).

2 Filonenko, N. N., R. Z. Sagdeev and G. M. Zaslavsky, *Nucl. Fusion*, **7**, 253 (1967).

3 Cary, J. R. and R. G. Littlejohn, *Annals of Physics*, **151**, 1 (1986).

4   Elsasser, K., *Plasma Phys. and Contr. Fusion*, **28**, 1743 (1986).
5   Arnold, V. I., *Mathematical Methods in Classical Mechanics* (Springer, New York, 1978).
6   Moffatt, H. K., *Magnetic Field Generation in Electrically Conducting Fluids* (Cambridge University Press, Cambridge, 1978).
7   Zel'dovich, Ya. B., A. A. Ruzmaikin and D. D. Sokolov, Magnetic fields in astrophysics. In: *The Fluid Mechanics of Astrophysics and Geophysics*, Vol. 3, ed. P. H. Roberts (Gordon and Breach, 1983).
8   Aref, H., *J. Fluid Mech.*, **143**, 1 (1984).
9   Kharhar, D. V., H. Rising and J. M. Ottino, *J. Fluid Mech.*, **172**, 419 (1986).
10 Ottino, J. M., G. M. Leong, H. Rising and P. D. Swanson, *Nature*, **333**, 419 (1988).
11 Arnold, V. I. *Compt. Rend. Acad. Sci., Paris*, **261**, 17 (1965).
12 Henon, M. *Compt. Rend. Acad. Sci., Paris*, **262**, 312 (1966).
13 Dombre, T., U. Frisch, J. M. Green, M. Henon, A. Mehr and A. M. Soward, *J. Fluid Mech.*, **167**, 353 (1986).
14 Childress, S., *J. Math. Phys.*, **11**, 3063 (1970).
15 Arnold, V. I. and E. I. Korkina, *Vestnik Moscow Univ., Matem. and Mekh.*, **3**, 43 (1983).
16 Galloway, D. J. and U. Frisch, *Geophys. Astrophys. Fluid Dynamics*, **29**, 13 (1984).
17 Galloway, D. I. and U. Frisch, *Geophys. Astrophys. Fluid Dynamics*, **36**, 53 (1986).
18 Finn, I. M. and E. Ott, *Phys. Rev. Lett.*, **60**, 760 (1988).
19 Galloway, D. I. and U. Frisch, *The Hydrodynamic Stability of the ABC-flows* (Nice Observatory, 1985).
20 Chernikov, A. A., R. Z. Sagdeev, D. A. Usikov and G. M. Zaslavsky, Weak chaos and structures, *Sov. Sci. Rev. C. Math. Phys.*, **8**, 21 (1989).
21 Zaslavsky, G. M., R. Z. Sagdeev and A. A. Chernikov, *Zh. Eksp. Teor. Fiz.*, **94**, 102 (1988).
22 Beloshapkin, V. V., A. A. Chernikov, M. Ya. Natenzon, B. A. Petrovichev, R. Z. Sagdeev and G. M. Zaslavsky, *Nature*, **337**, 113 (1989).
23 Arter, W., *Phys. Lett.* **97A**, 171 (1983).
24 Palm, E., *J. Fluid Mech.*, **1**, 183 (1960).
25 Jenkins, D. R., *J. Fluid Mech.*, **190**, 451 (1988).
26 Stuart, J. T., *J. Fluid Mech.*, **18**, 481 (1964).

CHAPTER 10

1 Wade, D., *Pattern in Islamic Art* (Studio Vista, a division of Cassell Publishers plc, London, 1976).

2 Bulatov, M. S., *Geometrical Harmonization in Architecture of Middle Asia in IX–XV C.* (Nauka, Moscow, 1988).

3 Hrbas, M. and E. Knobloch, *The Art of Central Asia* (P. Hamlyn, London, 1965). Copyright is held by Artia Foreign Trade Corporation, Prague.

4 Grunbaum, B. and G. C. Shepard, *Tilings and Patterns* (W. H. Freeman, New York, 1987).

5 Critchlow, K., *Islamic Patterns* (Thames and Hudson, London, 1976).

6 Zaslavsky, G. M., R. Z. Sagdeev, D. A. Usikov and A. A. Chernikov, *Usp. Fiz. Nauk*, **156**, 193 (1988).

7 *Patterns of Symmetry*, ed. M. Senechal and G. Fleck (University of Massachusetts Press, Amherst, 1977).

8 Weyl, H., *Symmetry* (Princeton University Press, Princeton, 1952).

9 *The World of M. C. Escher*, ed. J. L. Locher (Abrahams, New York, 1971).

10 Escher, M. C., *Art and Science*, ed. H. Coxeter *et al.* (North-Holland, Amsterdam, 1987).

11 Schattschneider, D., In [10], p. 82.

12 Schrödinger, E. *What is Life?* (Cambridge University Press, Cambridge, 1944).

13 Belov, N. V., *Notes on Structural Mineralogy* (Nauka, Moscow, 1976).

14 Bunn, C., *Crystals, Their Role in Nature and in Science* (Academic Press, New York and London, 1964).

15 Haeckel, E., *Kunstformen der Natur* (Bibliographisches Inst. Leipzig–Wien, 1899).

16 Bonnet, C., *D'Histoire Naturelle et de Philosophie*, V. 4, Recherches sur l'usage des feuilles dans des plantes, ed. C. S. Fauche (Librairie du Roi, 1779).

17 Rivier, N. *J. de Physique*, **47**, Coll. C3, Suppl. 7, 299 (1986).

18 Thompson D'Arcy, W., *On Growth and Form* (Cambridge University Press, Cambridge, 1942).

19 Cook, T. A., *The Curves of Life* (Constable, London, 1914).

20 Bravais, A. and L. Bravais, *Ueber die Geometrische Anordnung der Blätter und der Blüthenstande* (Breslau, 1839).

21 Church, A. H., *On the Relations of Phillotaxis to Mechanical Laws* (Williams and Borgate, London, 1901–1904).

22 Coxeter, H. S. M., *Introduction to Geometry* (Wiley, New York, 1965).
23 Gardner, M., *Scient. Amer.*, **236**, 110 (1977).
24 Stevens, P. S., *Handbook of Regular Patterns* (MIT Press, Cambridge, 1980).

Index

Printed in the United States
By Bookmasters